Real Process Improvement
Using the
CMMI®

Other CRC/Auerbach Publications in Software Development, Software Engineering, and Project Management

The Complete Project Management Office Handbook
Gerard M. Hill
0-8493-2173-5

Complex IT Project Management: 16 Steps to Success
Peter Schulte
0-8493-1932-3

Creating Components: Object Oriented, Concurrent, and Distributed Computing in Java
Charles W. Kann
0-8493-1499-2

Dynamic Software Development: Manging Projects in Flux
Timothy Wells
0-8493-129-2

The Hands-On Project Office: Guaranteeing ROI and On-Time Delivery
Richard M. Kesner
0-8493-1991-9

Interpreting the CMMI®: A Process Improvement Approach
Margaret Kulpa and Kent Johnson
0-8493-1654-5

Introduction to Software Engineering
Ronald Leach
0-8493-1445-3

ISO 9001:2000 for Software and Systems Providers: An Engineering Approach
Robert Bamford and William John Deibler II
0-8493-2063-1

The Laws of Software Process: A New Model for the Production and Management of Software
Phillip G. Armour
0-8493-1489-5

Real Process Improvement Using the CMMI®
Michael West
0-8493-2109-3

Six Sigma Software Development
Christine Tanytor
0-8493-1193-4

Software Architecture Design Patterns in Java
Partha Kuchana
0-8493-2142-5

Software Configuration Management
Jessica Keyes
0-8493-1976-5

Software Engineering for Image Processing
Phillip A. Laplante
0-8493-1376-7

Software Engineering Handbook
Jessica Keyes
0-8493-1479-8

Software Engineering Measurement
John C. Munson
0-8493-1503-4

Software Engineering Processes: Principles and Applications
Yinxu Wang, Graham King, and Saba Zamir
0-8493-2366-5

Software Metrics: A Guide to Planning, Analysis, and Application
C.R. Pandian
0-8493-1661-8

Software Testing: A Craftsman's Approach, 2e
Paul C. Jorgensen
0-8493-0809-7

Software Testing and Continuous Quality Improvement, Second Edition
William E. Lewis
0-8493-2524-2

IS Management Handbook, 8th Edition
Carol V. Brown and Heikki Topi, Editors
0-8493-1595-9

Lightweight Enterprise Architectures
Fenix Theuerkorn
0-9493-2114-X

AUERBACH PUBLICATIONS

www.auerbach-publications.com
To Order Call: 1-800-272-7737 • Fax: 1-800-374-3401
E-mail: orders@crcpress.com

Real Process Improvement
Using the
CMMI®

Michael West

Foreword by
Barry Boehm

AUERBACH PUBLICATIONS

A CRC Press Company
Boca Raton   London   New York   Washington, D.C.

Library of Congress Cataloging-in-Publication Data

West, Michael.
    Real process improvement using the CMMI® / Michael West.
      p. cm.
    Includes bibliographical references and index.
    ISBN 0-8493-2109-3 (alk. paper)
   1. Production management. 2. Quality control. 3. Capability maturity model (Computer software) I. Title.

TS155.W459 2004
658.5—dc22

2003070899

Visit the Auerbach Web site at www.auerbach-publications.com

© 2004 by CRC Press LLC
Auerbach is an imprint of CRC Press LLC

No claim to original U.S. Government works
International Standard Book Number 0-8493-2109-3
Library of Congress Card Number 2003070899
Printed in the United States of America 3 4 5 6 7 8 9 0
Printed on acid-free paper

DEDICATION

This book is dedicated to my wife Jitka who for so many years has patiently taught me and encouraged me to believe in myself. It is also dedicated to my friend and business partner Donna Voight who has inspired me to continuously look inside myself and find improvement. I am grateful to Gene Miluk at the SEI for being my coach and mentor in organizational change. I thank all of Natural SPI's clients for the most precious learning opportunities, many of which have gone into this book. Finally, I offer a special thank you to Barry Boehm, one of my heroes, for his review of the manuscript and for providing a fitting Foreword and to Tom Waters who developed the book's illustrations and thus gave it the professional look it deserved.

TABLE OF CONTENTS

FOREWORD

You've probably heard about the South Sea Island cargo cults, but let me refresh your memory a bit. Some good examples happened during World War II when military people set up supply chain airbases on some of the islands. The native people watched the military people set up radars and radio telephone boxes and phones. Then the military people would talk into the phones and airplanes would fly in with cargo, which was shared with the native people.

When World War II ended, the military people packed up their equipment and left and no more airplanes or cargo came. So the native people attempted to bring the airplanes and cargo back by setting up wooden radars, boxes, and phones, dressing up in uniforms, and shouting into the phones. But no cargo came, and they were mystified and frustrated.

Unfortunately, the same cargo cult syndrome can and often does happen with Capability Maturity Models (CMMs), including the CMMI®. CMMs capture processes used by successful organizations and surround them with such modern cult appurtenances as trademarks, service marks, and licensed assessors. Unsuccessful organizations that want to develop on-time, within-budget, high-quality software and systems often set up these processes and expect that by following rigorous procedures for requirements management, planning and control, and quality assurance, that on-time, within-budget, high-quality systems will emerge for them too. But all too often, very little useful "cargo" appears and the organizations are mystified and frustrated.

In this book, Michael West does a great service for many potential CMM or CMMI adopters by explaining the differences between pro forma and in-depth organizational process improvement. He also illustrates the differences through many examples from hard-won personal experience as a process group member and process improvement consultant for a wide variety of organizations. He shows the difference between attacking symptoms (e.g., trying to use requirements management change control procedures to eliminate scope creep) and addressing root causes (e.g., poor customer communication, overinterpreting "the customer is always right," lack of change impact analysis). He debunks various myths about CMMs (e.g., higher maturity levels will fix your problems and guarantee success, organizations operating without CMMs are in total chaos, CMMs are expert-only "rocket science"), and provides case studies from experience showing that the myths are dangerous to believe.

More than this though, he provides positive guidance on how organizations can draw on the CMMI or other CMMs to effect real process improvement. He emphasizes that most organizations have some highly effective practices that fit their culture and business environment and shows that these need to be understood, conserved, and built upon rather than being trashed in a "slash and burn" approach to process improvement.

He provides a good project-oriented sequence of steps for addressing the root causes rather than the symptoms: establishing a common language, determining your organization's starting point, establishing goals and success criteria, planning the improvement project, and monitoring and controlling its progress.

Each chapter has some good features for keeping the reader oriented and continuously engaged, starting with a "What Do You Think? What Do You Believe?" assessment, summarizing what to do and what not to do, and ending with a "What Did You Learn? What Will You Do?" assessment.

If you're contemplating any sort of process improvement effort, this book can help you a great deal in defining and getting the results you'd like to achieve.

Dr. Barry Boehm
TRW Professor of Software Engineering,
Computer Science Department
Director, USC Center for Software Engineering

INTRODUCTION

WHAT YOU'LL MISS IF YOU DON'T READ THIS INTRODUCTION

Do most people read a book's foreword, prologue, or introduction? I haven't always, but I do now because I found out what I was missing. This introduction will give you some very specific and useful instructions which will enable you to make the most use of the book. Unlike other tomes, you do not have to read this book cover to cover to learn something and benefit. In fact, you may benefit more quickly from the information if you skip around and just read the chapters or sections that contain information you need right now.

Those of you who are savvy in the ways of project management will get this analogy: When I began writing this book in earnest, I viewed it as a project. So the first thing I needed to do was to scope out the project. I needed to define — at least at a high level — what goals the book needed to accomplish or what problems it needed to solve. I also needed to figure out the approach the book would take to accomplish its goals. In other words, I needed a project charter.

This introduction is that project charter. It describes the project's background and how and why this book has come into existence. It tells you what you can expect to get out of reading the book and it points you to the specific chapters or sections that give you information you need to apply in your particular environment. It also gives you an overview of some of the book's more critical design structures, so you can use the book more effectively and efficiently. It will be nice when people read this book, but it will be far more rewarding when they *use* it.

THE LONG, HARD ROAD TO HERE

My career in systems process improvement (SPI) began in June of 1996 when I took a new position within Xerox as a member of the Software Engineering Process Group (SEPG) in the company's Production Systems Group (PSG). Back then, it was just software process improvement because it was based on Software CMM® and CMMI was maybe only a twinkle in Mike Konrad's eye. I left Central Florida, my home of 36 years, and moved to El Segundo, California — process that is… KPAs… no movie stars.

My introduction to the Capability Maturity Model® for Software (SW-CMM®) was magical: a personal and professional epiphany. In this simple model for process improvement, I saw the cure to all that ailed software development, the be all and end all. I very quickly became an apostle of process improvement via implementation of CMM. In my position as de facto spokesperson for the division's SEPG, I missed no opportunity to preach the gospel of the five maturity levels and their blessed key process areas (KPAs). In the business days before it was chic to make fun of those who believed in holy grails, silver bullets, and pixie dust, I was convinced CMM was all three rolled up into one. So were a lot of other people. Of course, we were mistaken.

Less than four years later and still working in process improvement, I began turning around inside. Secretly and inwardly, I felt myself about to become one of the most vocal and antagonistic critics of the software process improvement initiative in Xerox. Thinking that perhaps the cause of my angst was peculiar to the Xerox environment (which is peculiar), I went to work for Computer Sciences Corporation (CSC) with the belief that they were doing things right with CMM. It took less than two years for me to find out they were not, and again I become disillusioned. Naïve me.

After considerable introspection, I finally came to grips with what had happened to me: the former CMM evangelist, while still completely enamored with the beauty of the model itself, had become quite disillusioned with the way in which most organizations were implementing CMM. By this time, many more organizations were making the same implementation mistakes with CMMI. I also came to the realization that so many people focus so much on the CMM/CMMI models that they lose sight of process improvement.

Thus this book. Sure, I could spend hundreds of pages lambasting management ineptitude in implementing process improvement in software and systems organizations. But then this book would be just one more of many in the useless genre of "I told you so." And, hey, by the way, I was one of those inept managers. I'd rather not focus on what people have done wrong and continue doing wrong. Instead, my contribution to people trying to make CMM or CMMI work for them is to share some valuable lessons I and others have learned in the hope that such lessons can be used to enable a smoother, easier, and quicker approach to process improvement.

I think the biggest problem people have with implementing Software CMM and CMMI is that they tend to view the model as something foreign or alien to software and systems development. Yet, when you really come to know these models and how they can be implemented, you begin to see just how naturally the models align with the way you would develop and deliver software and systems if you had the freedom to do things the "right" way. Despite what their critics believe, CMM and CMMI really are intuitive and natural. They represent the way people in software and systems organizations already work or want to work.

To prove this point, conduct this little experiment. Go out and ask some engineers, integrators, and project managers in your organization to describe how things would work if they were the bosses. Not surprisingly, you'll get original and organic answers from people that describe processes that can be closely mapped to SW-CMM and CMMI.

You'll get answers from engineers like, "I wish the customers would make up their minds what they want." Translation: Define the requirements and manage them. Project managers/leads will say, "I'd like for my senior managers to stop changing the priorities every day." Translation: Manage (e.g., define, document, approve, and communicate) project commitments and find a way to keep them.

This is not coincidence. As cited in Software CMM's[1] Chapter 1, Section 1.1, "The Evolution of the CMM," SW-CMM is based on actual software practices and knowledge gained from the study of successful software organizations. CMMI[2] improves on the CMM by incorporating real-world lessons learned from people who used CMM. CMMI does not introduce anything new to systems development and delivery, just as Software CMM didn't introduce anything revolutionary to software development. Both models simply codify what successful software and systems delivery organizations have been doing for many years. Translation: As a matter of practicality, if SW-CMM and CMMI are collections of good software/systems engineering and management practices, then these models would exist whether or not they were ever published as books.

Confounding the pervasive misunderstanding of these models is that people want to view them as prescriptive. It seems that no matter with whom I speak in the process improvement community, they seem to want to read more into the models than what is there. Primarily, people trying to implement CMM or CMMI want to believe that it says how to do things, when, in fact, the model merely addresses what successful organizations do. People use meaningless phrases like "SEI^{SM} requirements" or "CMMI requirements" when, in actually, nothing of the sort exists.

Thus the crux of this book: CMMI defines the processes that most reasonable people in systems development and delivery would do naturally if allowed to do so. If you really want to implement CMMI in your organization, don't! Don't try to force fit an academic model to your business. Instead, find what your business is already doing well and what it needs to do better. Map those good practices to the model and find what things in the model address what your organization is not doing so well. This book is not so much about CMM or CMMI as it is about process improvement; as you will see, the two are not synonymous.

Process improvement is inseparable from organizational change because introducing process or process improvement is change. There are dozens of books on the market — some of them referenced herein — which explore and explain managing organizational change and this book doesn't attempt to repeat those good works. However, because you cannot do process improvement without change considerations, there are digestible bits and pieces of practical organizational change advice interspersed throughout the text. Instead of being generic, this advice on change is specifically integrated with the "how to" text on process improvement to maximize its usefulness to you.

Another key message in this book is this: everything in CMMI may represent process improvement, but not all process improvement is in CMMI. CMMI is just one subset of the larger, more universal superset of things you can do to improve your software/systems delivery. It is very much like the field of astrophysics: each time scientists think they've discovered the boundaries, they soon discover that the known is just a small part of a much larger, grander thing. CMM or CMMI is just a small part of the greater and grander universe of process improvement. For more information on this idea, check out "CMMI's Place in the Process Improvement Universe" in Chapter 1 — News Flash! There Is a Level 1.

So the problem with focusing strictly on implementing CMMI (which you cannot do anyway) is that it becomes too easy to lose sight of many other opportunities for process improvement. People sometimes believe that if only they implement CMMI, they'll get process improvement. Many of the concepts and experiences described in this book suggest that the opposite is more likely: if you build a culture for process improvement, CMMI and maybe even maturity levels will come to your organization as a natural by-product.

Now back to my personal journey in process improvement. After leaving CSC and having over 20 years in corporate America learning mostly how things should not be done, a business partner and I struck out on our own and formed the process improvement and management consulting firm, Natural Systems Process Improvement (Natural SPI, Inc.). We founded our consulting practice on some very unconventional concepts and approaches to CMMI-based process improvement. So far, our unique message has been warmly welcomed by clients and the results of our work have genuinely benefited our client organizations. This book describes most of Natural SPI's successful concepts, approaches, and implementation practices.

Another pervasive theme you'll find throughout this book is that I try to balance thinking with acting and strategy with execution. On the Birkman personality profile, I'm a documented freak: I think both globally and linearly, equally. Thinking great thoughts (which I don't claim to do) without being able to apply them is, for me, like fantasizing or daydreaming; it's fun and may be good for me, but not very useful for anyone else. Acting without first thinking, executing without a plan, is almost always just plain stupid. So this book gives you the science and the technology, the proven concepts, and the practical steps for applying them.

I encourage you to have as much fun reading this book as I have had writing it. To illustrate concepts, I often point to mistakes people make in process improvement. I know of what I write for I have made or suffered every one of these mistakes. I made the mistakes, had a good laugh at myself, learned from them, and moved on. These mistakes are expensive. If your organization can avoid even one of the mistakes I address in this book, you've gotten your purchase price returned many times over.

WHY THIS BOOK?

There are already numerous books about CMMI on the market from which to choose, so why choose this one? What sets it apart from other books? Here's what you'll find different in these pages:

There are many myths about CMMI and its implementation. Some books perpetuate these myths. This book debunks and destroys process improvement myths that too often inhibit people's work and progress. This book exposes CMMI myths by shining the harsh light of fact on them. People who benefit from perpetuating those myths are going to be very angry with me. Ah, sweet success.

I eventually pulled my head out of my process; you can too! And when you do, you'll discover, as I did, that there are many industries and disciplines from which you can borrow good concepts and practices and apply them to your CMMI process improvement work. The Software Engineering Institute (SEI) didn't invent process improvement, and it's folly to think that software and systems engineering process improvement wouldn't exist without CMMI. This book tells you why, what, and how.

This book in no way pretends to have all the answers or solutions; no book can, despite the claims made by its author or publisher. But what this book will do, more so than any other, is help you learn how to formulate the important questions about process improvement that will then lead you to the answers. Questions are good, and this book is chock full of them.

GOALS FOR THIS BOOK

This book is written to provide you with some very specific, practical knowledge about implementing CMMI-based process improvement in your organization. The knowledge you can expect to gain includes:

■ A thorough debunking of any beliefs or thoughts you might have had that organizations which don't use CMMI or haven't been appraised at a CMMI maturity level are in chaos, not doing anything right, and are condemned to fail.
■ An understanding that if the principles and practices of project management work for software and systems projects, they'll also work for process improvement projects.
■ Unconventional, yet highly effective methods for accomplishing the activities that usually bog down every CMMI effort: defining, developing, piloting, and implementing the process assets.
■ Insight into some different ways of looking at CMMI-based process improvement that may give you your own breakthrough thoughts on how to positively effect this type of change in your organization.
■ Some fresh approaches to some of the most difficult areas of CMMI-based process improvement: designing and defining processes and process assets.
■ A special chapter just for executives and senior managers from which they can learn explicit techniques to implement and become effective leaders and sponsors of process improvement.
■ Questions, criteria, and other aspects of critical thinking your organization should employ when considering acquiring outside CMMI or process expertise or purchasing a tool or system to be used to enable process implementation.
■ An understanding of the myths that pervade CMMI-based process improvement, the dangers of myths, and how to separate fact from fiction.

So, for those of you who have a penchant for quality assurance, please use the above for goals to build your checklist and then see if this book passes your audit!

What This Book Will Do for You

Specifically, what you will get from using the contents of this book is sound, proven, practical advice on how to use CMMI (and other tools) to improve your organization's software and systems development and delivery processes. You will be able to immediately apply the concepts and put the practices into operation in your organization. The results of your work will not be a surprise; you will be able to measurably demonstrate real improvement in your organization.

What This Book Will Not Do for You

There are a few things this book will not help you with. If you've already bought the book and want to do any of the following, Amazon.com will help you sell it as used so you can recover some of your cost. This book will not:

- Give you anything that guarantees your organization will achieve a maturity level, because no one or nothing can deliver on that promise. However, there are plenty of people out there who will happily relieve you of your money if you believe this.
- Give you only one way to implement engineering and management practices consistent with CMMI, because there are an infinite number of ways to do this. It will give you several ways to accomplish your goals, which have proven successful in other organizations.
- Give you an inexpensive, fast-track approach to improving process capability or organizational maturity. No matter what anyone says, this stuff is hard and expensive. This book will give you approaches that are less painful and less expensive.

WHO CAN BENEFIT FROM THIS BOOK (AND WHO CANNOT!)

People in almost every role or function in any systems delivery or support organization in the world can benefit from one or more of the fresh ideas and practices explained in this book. Executives and senior managers, program and project managers, engineers, designers, architects, testers, and especially people with process focus responsibility, such as SEPG members or process improvement managers or leads, will all benefit from various chapters and sections.

If you want to improve the way your organization works; if you want people to work more effectively and efficiently; if you want your organization to stop wasting time, money, and energy on activities that don't add value to anything, then this book is for you.

If, however, you want to achieve a CMMI maturity level and don't really care about improving processes or if you want to just talk about improving your organization but don't really want to do anything, this book is definitely not for you.

Assumptions about You, the Reader

In writing this book, I have made certain assumptions about the reader's knowledge and experience. Out of necessity — and because I did not want to repeat the entire CMMI — I assume that the reader has a basic understanding of CMMI, including knowledge of the process areas and the acronyms by which they are commonly known (e.g., REQM, PMC, OT, OPP).

I also assume the reader has at least primitive knowledge or awareness of some basic project management concepts, such as planning, estimating, and reporting progress and status. Finally, I assume the reader is capable of invoking some level of critical thinking. Like every book about CMM or CMMI, including the models themselves, you should not blindly accept the ideas and practices contained in these pages. Thinking critically and independently is hard work, requires you to have courage, and is essential to your success in using this book.

HOW TO USE THIS BOOK

Whether you're reading this text online or in hard-copy form, you should feel free to consume the information any way you like. As much as possible, I've tried to make each chapter or section a self-contained unit of information so you can get from it the essence of the knowledge it's trying to convey independent of other chapters and sections. Naturally, I've arranged the information in a sequence that seems logical to my own mind, but if you talk to my wife or friends who know my mind, you may not want to take that route.

This book is, by no means, the most comprehensive word on model-based process improvement and there are other texts, many of them referenced herein, which you should also read.

The information in this book is organized into eight chapters:

1. News Flash! There Is a Level 1!
2. The Role of Roles
3. Managing the Process Improvement Project
4. Process Improvement Strategies that Work
5. Five Critical Factors in Successful Process Definition

6. Acquiring Process Expertise and Tools
7. Effective Change Leadership for Process Improvement
8. Process Improvement Myths and Methodologies

Which Parts You Should Read

In any organization, the functional roles of people who are directly involved in CMMI-based process improvement can fit into one of the following four general categories:

1. Executives and senior managers
2. People with process focus responsibilities, such as SEPG members or process improvement managers
3. System engineering program or project managers
4. Engineers and users of organizational processes

Most of the information contained in this book is of greatest value to process improvement leads or managers and members of groups responsible for some aspect of CMMI-based process improvement, such as SEPG members.

However, some sections are more relevant and useful to people in one of these four categories than to others. In these instances, you'll find the text is preceded by a picture that indicates the group to which that section is targeted. This makes it easy for anyone in the organization to scan the book and find the information they need most. Of course, I would be very happy if everyone read everything, but I'd rather help you get what you need when you need it.

Executive and Senior Managers

 If you're a senior or executive level manager, you may have numerous projects, programs, and initiatives to oversee and manage. You're interested in high-level information because there's just too much detail and, besides, you've got people reporting to you whom you're counting on to know and watch the details.

The information contained in these sections primarily focuses on what is happening and why you should care. These sections will give you enough information to be able to ask the right questions and make the right decisions. There is also a special chapter devoted to you titled, "Effective Change Leadership for Process Improvement." If you don't read anything else, you need to read this chapter if you want your organization's CMMI or process improvement effort to do more than just spend money and use valuable resources.

Systems Project Managers and Leads

CMMI-based process improvement simply will not succeed without your involvement. Without software or systems engineering projects, there wouldn't even be much need for CMMI. If your organization's senior management is going to push CMMI on the organization, you might as well get involved in the effort early and as much as possible. You can and should influence the results and outcomes of the improvement efforts, but you won't be able to if you try to avoid it.

These sections give you what you need to not be steamrolled by those "process people." You'll not only learn how to cope with the changes being made, you'll learn how to make the process improvements work to the benefit of you, your projects, and your customers.

Engineers and Other Process Stakeholders

You are the people often called the "victims." You're the people who are using new or revised procedures. You are project technical leads, engineers, architects, developers, testers, integrators, suppliers, and people responsible for configuration management or change control. As far as you're concerned, you are the real people with the real jobs. You might be people who would very much like for all this process stuff to go away.

Since it's not going to go away, these sections are written to give you the minimum amount of information you need, without annoying you with a lot of stuff you don't care about. Just try it; it won't hurt that much.

SPECIAL SECTIONS

Throughout the book, you'll find special sections that are designed to help you understand, retain, and use the information being imparted. Some special sections will also help you have some fun learning the information, while others will hopefully cause your mind to just wander off down its own path so you can make new discoveries and explore breakthrough ideas of your own. These special sections are described below.

Quiz: What Do You Think? What Do You Believe?

At the beginning of each chapter, there is a short quiz. I can't watch you use this book, so I don't know if you'll take the quiz or not, answer the questions honestly or cheat. However, I think if you just try one, you'll want to take them all because you'll have fun.

The purpose of having a quiz at the beginning of each chapter is to help you find out how much you already know about the topic of discussion in that chapter. These quizzes also help you discover preconceptions, prejudices, and myths about the topic that you may have been carrying around and using as if they were truths or facts.

The answers to the quiz questions are not handed to you, but they're not hard to find either. Even a cursory reading of the chapter will render the correct answers.

Quiz: What Did You Learn? What Will You Do?

At the end of each chapter, there is another chance for you to participate in your own learning. This section challenges you to take what you've learned and immediately put your newfound knowledge into a plan of action, which you can then integrate with your planned or on-going process improvement work. This section will help you formalize and document your thoughts about what to do next and how, for whom, and with whom.

Natural Change

 There are many fine references on the market today that address almost every known aspect of organizational and cultural change. This book makes no attempt to repeat, rehash, or improve on these works. Process improvement, whether it's via the implementation of standards such as ISO, using CMMI as the basis for improvement, or Six Sigma, represents a particular flavor of organizational change. And because I believe changing corporate culture is more effective — certainly more efficient — when it starts at the top, I've included a chapter on this topic that will primarily benefit the executive or senior management sponsors of process improvement.

Yet, there are many gems of practical knowledge and executable truths pertinent to CMMI-based process improvement which I and my associates have come across. I share those with you throughout the text in sections with this title: Natural Change. The purpose of these golden snippets of information is to get you to stop and think about your current CMMI work and to look at your efforts through different and insightful perspectives.

Do's and Don'ts

There's a certain percentage of you who love checklists, so checklists it is! At the end of each chapter, you'll find the most critical, pertinent lessons from that chapter summarized in a list of things you should do and things you should not do. Yes, of course you can skip the chapter and go straight to the checklist, but the checklist alone will not give you the full benefit of the concepts and actionable information found in the chapter. So be careful not to cheat yourself.

WHAT DO YOU WANT TO DO RIGHT NOW?

Okay, now if you've been patient enough to read through this introduction, you're going to leap ahead of those who skip it and jump right into the chapter because Table I-1 serves as a roadmap for getting you directly to the information you want or need right now.

Scan the left column of Table I.1 for topics that represent something you want to do or know. Then, follow that row across to the right column, which tells you what you need to learn and where in the book you need to look to find that information.

Table I.1 Where to Go

What is your question? What do you want to achieve?	Go to ...
My organization is new to CMMI and process improvement. Where do we start?	Chapter 3 — Managing the Process Improvement Project will help you get started setting process improvement goals. However, since you'll need to know the organization's true starting point, this path will eventually lead you to Chapter 1 — News Flash! There Is a Level 1.
We're using CMMI, but we don't seem to be making much progress. What should we do?	Check out Chapter 4 — Process Improvement Strategies that Work for some different approaches to CMMI-based improvement.
We're never sure who is supposed to do what. How do we figure out what people's jobs are in relation to process improvement?	Defining roles and responsibilities is one of the most underprioritized, yet critical prerequisites to successful CMMI implementation. Read Chapter 2 — The Role of Roles.
We're thinking about hiring a CMMI consultant to help us out. How can we hire the right people?	Chapter 6 — Acquiring Process Expertise and Tools can help your organization establish fact-based rationale for bringing in outside expertise and for selecting a consultant that best fits your organization's needs.
People say lots of things about CMMI, but some of it just doesn't sound right. What's true and what's not true about CMMI?	The only thing that is true is what you believe is true. So the real decision is whether you want to base your beliefs on myths or on facts. Take a look at Chapter 8 — Process Improvement Myths and Methodologies to see of what you're hearing about CMMI-based process improvement is real.

(continued)

Table I.1   Where to Go (Continued)

What is your question? What do you want to achieve?	Go to ...
Our organization hasn't yet had a maturity appraisal. Does that mean we're at Maturity Level 0 or 1?	Not necessarily. Your organization could be operating at a CMMI Maturity Level 5, but it just hasn't been measured yet. You wouldn't have to climb Mount Everest to know that it's taller than ten feet. But this isn't really the point. Many organizations are already doing things which they can leverage and reuse in a CMMI-based improvement effort. Go to Chapter 1 — News Flash! There Is a Level 1!
Creating processes and procedures doesn't seem to be very structured. Are there some lessons to be learned in process definition?	Yes. Chapter 5 — Five Critical Factors in Successful Process Definition will give you some proven approaches and techniques in this topic which you can apply immediately in your organization's CMMI work.
My management disgusts me! They say they want a maturity level, but they don't really support the process improvement effort. Any ideas?	Yep. Slip this book under their door with a clip marking Chapter 7 — Effective Change Leadership for Process Improvement. The whole chapter is less than 20 pages long, so even a busy executive with a short attention span can read it.
Any advice on buying a tool to help us create and deploy our processes and procedures?	Yes, don't let anyone sign the purchase order until all the decision makers have read Chapter 6 — Acquiring Process Expertise and Tools.
We don't want to repeat the mistakes others have made in CMMI implementation.	Good for you! Read the whole book. It is replete with costly mistakes made by other organizations and how to avoid making them again.

1

NEWS FLASH! THERE IS A LEVEL 1!

The map is not the territory.

— **Alfred Korzbyski**

QUIZ: WHAT DO YOU THINK? WHAT DO YOU BELIEVE?

Take a minute and answer the questions to the quiz in Figure 1.1. Then, once you've finished reading this chapter, take the quiz in Figure 1.7 — "What Did You Learn? What Will You Do" — to find out how much this information has helped you with your own CMMI-based process improvement.

As you read this chapter, you'll discover that definitive answers to questions about CMMI-based process improvement can be evasive. When asked questions, seasoned instructors at SEI have learned to answer with "it depends" or "you decide" and they are wise to do so. When it comes to cultural or technological change, which encompasses process improvement, the "right" answer is the one that most benefits your organization and its business needs.

THE MODEL AND THE REALITY

The Staged Representation of CMMI identifies five levels of capability maturity and defines process areas (PAs) for Maturity Levels 2 through 5. Why are there no PAs in Level 1? And if there are no PAs in Level 1, why is it called Level 1 and not Level 0 (zero), and why doesn't the maturity scale go from zero to four instead of one to five? And can an organization really make the quantum leap from mastering no practices or process areas to many?

1. **An organization that has not been appraised at a CMMI maturity level:**
   a. Is in total chaos
   b. Has no processes
   c. Cannot deliver on schedule and within budget
   d. Has happier employees
   e. Not necessarily any of the above

2. **True or False:** The best way to implement process discipline is to write new procedures and make people use them.

3. **Process improvement includes which of the following:**
   a. Managing requirements
   b. Making meetings more efficient
   c. Planning and managing projects
   d. Improving communication
   e. Some of the above
   f. All of the above

4. **True or False:** Your organization has to approach the CMMI differently because your organization is different from others.

5. **The best goal for CMMI-based process improvement is** _____ .

Figure 1.1   Chapter 1: What Do You Think? What Do You Believe?

If you read nothing else in this entire book, please read the section in this chapter titled, "Slash-and-Burn versus the Natural Approach to Process Improvement," for it encapsulates the core ideology of this book in a few short paragraphs.

As an executive or senior manager of a software or systems organization undergoing CMMI-based process improvement, you need to be aware of and supportive of the fact that people in your organization have already developed good engineering and management practices. Do not automatically assume that people need to change what they're already doing. Encourage your process focus people (SEPG, EPG, or whatever they call themselves) to find ways to leverage the existing best practices and procedures in introducing CMMI-based process improvements.

An effective way to determine your organization's current process capability or organizational maturity is to conduct some type of process appraisal such as a Standard Capability Appraisal Method for Process Improvement (SCAMPI^SM).[3] However, the type (and relative cost) of appraisal you conduct to figure out your organization's starting point for process improvement should be based on business decisions and an understanding of the organization's recent history. Don't let this important decision be based on what some consultant wants to sell you. For more information on appraisals and figuring out the starting point, read the section, "Determining the Starting Point for CMMI," in Chapter 2 — Managing the Process Improvement Project.

Finally, since the success of your organization's CMMI effort is critically dependent on your leadership, you should read Chapter 7 — Effective Change Leadership for Process Improvement.

Very soon after you hear that your organization is going to do some process improvement or CMM or CMMI stuff, the people responsible for process in your organization may come around asking people like you questions about how you do your work. What they're trying to do is find out what the organization is doing well and what can be improved in terms of software development and management processes. (If no one comes around asking such questions, find out why not.)

It is very much to your benefit to be cooperative and help the process people gather this information and here's why: If the process people (i.e., the SEPG) cannot find out what procedures people are currently using to do their work, they might go off and invent procedures which you won't like but will have to follow. If you speak up and let them know what you're doing, how you're doing it, and that things are working just fine thank you, then the process people will use your existing procedures (the ones that you're used to and like) as the starting point. They may make some incremental changes to these procedures down the road, but your contributions to the start-up of the process improvement initiative will probably keep management from thinking that a radical change needs to occur.

The reality of life in modern business is that organizations don't really leap from zero to six or more PAs in a single bound and there really are process areas in a Level 1 mature organization. They are not defined in the CMMI, but they can be found in the larger universe of process improvement.

This situation — the concept of a Level 1 organization existing with Level 1 being undefined — always reminds me of the 1984 Rob Reiner film, *This Is Spinal Tap*.[4] There's a scene in the film in which the characters

Marty and Nigel are looking at Nigel's guitar collection. They get around to talking about Nigel's favorite amplifier and the dialog goes like this:

NIGEL: This is a top to a, you know, what we use on stage, but it's very ... very special because if you can see ...

MARTY: Yeah ...

NIGEL: ... the numbers all go to eleven. Look ... right across the board.

MARTY: Ahh ... oh, I see....

NIGEL: Eleven ... eleven ... eleven....

MARTY: ... and most of these amps go up to ten....

NIGEL: Exactly.

MARTY: Does that mean it's ... louder? Is it any louder?

NIGEL: Well, it's one louder, isn't it? It's not ten. You see, most ... most blokes, you know, will be playing at ten. You're on ten here ... all the way up ... all the way up....

MARTY: Yeah....

NIGEL: ... all the way up. You're on ten on your guitar ... where can you go from there? Where?

MARTY: I don't know....

NIGEL: Nowhere. Exactly. What we do is if we need that extra ... push over the cliff ... you know what we do?

MARTY: Put it up to eleven.

NIGEL: Eleven. Exactly. One louder.

MARTY: Why don't you just make ten louder and make ten be the top ... number ... and make that a little louder?

NIGEL: ... these go to eleven.

This wacky little scene mirrors the strangeness of the CMM and CMMI maturity levels. Why not make it four levels — one to four — and have process areas and practices defined for Level 1 and not have a Level 5? The only answer I've ever been able to deduce is "this one goes to five."

In reality, there is a Level 1; it's just not defined in CMM or CMMI. Business organizations can be observed to share common characteristics which may not be directly related to CMMI. Furthermore, not all organizations that have not been appraised at some defined maturity level are in the same state. Organizations which have not been appraised at a maturity level can vary widely in their respective process capabilities. This chapter takes a look at some of the properties — I'll call them evolved business practices (EBPs) — which organizations can possess and sometimes have institutionalized long before they ever consider taking on CMMI-based process improvement or being appraised against the model. The information in this chapter also looks at how some of these EBPs can serve as powerful boosters to launching your process improvement effort.

But first, let's take a look at the biggest mistake organizations make right after they first get the idea to implement process improvements based on CMMI. I call this really commonplace, really bad idea Slash-and-Burn.

SLASH-AND-BURN VERSUS NATURAL PROCESS IMPROVEMENT

With not so fond memories, I often recall my years growing up in Central Florida. It was the late 1970s and early 1980s and the region was undergoing explosive growth. Amusement parks, tourist attractions, hotels, golf courses, and housing developments popped up, it seemed, overnight. Orange groves, pasture land, and woodlands were bulldozed, burned, and paved over at a frightening rate that could never seem to satiate the endless stream of new residents and hoards of greedy developers. To save money, developers would employ slash-and-burn techniques and destroy every living thing on the targeted land. Later, after putting up the buildings or sculpting the golf course, they would plant smaller, younger versions of the exact same trees and shrubs that had previously lived there. Presumably, this replanting was done to give the developed area a "natural" look. Not surprisingly, even after decades of expensive grounds maintenance, the reengineered patches of "wild" Florida never achieved the level of ecological splendor that had been leveled to dirt and ash in a few days.

The Slash-and-Burn Approach

Unfortunately, like the Florida "developers," slash-and-burn is also an approach often used by well-meaning but misguided people involved in process improvement, and it's an approach particularly favored among IT outsourcing firms. In slash-and-burn process improvement, process zealots, whether they take the form of new outsourced management, outside consultants, or insiders fresh back from some process class at SEI, march into the organization like an occupying army. They may look at the organization's process forest, but they see only the tangle of the underbrush and not the naturally evolved beauty and maturity. Without slowing down to get a feel for the existing culture or to take a survey of the existing process landscape to find what good practices may inhabit this landscape, these process people quickly determine that there is no process or discipline, only chaos. Just as quickly, the process crusaders determine that "the primitives need CMMI and they need it now whether they like it or not." After all, "it's for their own good and they'll thank us later."

To be fair, it's not always the process shock-troops who make the call to bulldoze the existing culture. More often than not, these heavy-handed approaches are the result of a command-and-control structure up to the highest levels in the organization. The CEO mentions in passing one morning over coffee that perhaps "we could use this CMMI thing to make some improvements" and by the time the message has been amplified downward through the minions of underlings wanting to please the boss, the message has become, "you'd better get to Level 5 by next year or else!"

A slight variation of slash-and-burn SPI occurs sometimes when those of us in the process business simply forget that we exist to improve business and technical processes for the benefit of individuals and organizations. When this happens, "process" becomes the business. We forget that the project managers, engineers, configuration specialists, and senior managers are our clients and that we serve them, not the other way around.

Symptoms of the Slash-and-Burn Approach

The "ethnic cleansing" version of process improvement isn't always as easy to notice as you might think. If you suspect that the organization you're in may have experienced or is experiencing slash-and-burn process improvement, look around for these symptoms:

- You ask a process person (such as a SEPG member) who their client is and how they are serving the client and you get a funny, confused look that says, "what are you talking about?"
- You can't find anyone who can name one goal for the process improvement effort other than the achievement of a maturity level.

- There's a prevailing belief that no processes existed before the CMMI initiative began. (Partial answer to Quiz Question 1.)
- None of the members of SEPG or your process focus group have any roles or responsibilities other than process or CMM/CMMI implementation.
- No matter how people were doing things before, they're not allowed to do it that way anymore; they have to follow the new procedures. (Partial answer to Quiz Question 1.)
- No matter what someone is doing, when you ask him or her why they're doing it, they tell you, "because the process requires it."
- People with software delivery responsibilities can recite CMMI practices or the identifications and titles of their organization's policies and procedures.
- Estimates for process overhead in development projects exceed 15 percent of the projects' total effort.
- The volume of standards and procedures increases, while the quantity and quality of delivered products decreases.
- People use words such as "audit," "inspection," and "compliance."
- People refer to "CMMI requirements" or "SEI requirements."
- People quietly make jokes about the "process police" or the "process Gestapo."

Results from the Slash-and-Burn Approach

Both the slash-and-burn approach and the "process is the business" attitude can have unintended but disastrous consequences, including but not limited to:

- Soon (within six months) after the outsourcing cut-over or the initiation of the process improvement effort, the highly skilled or expert employees and managers leave the organization in barely-concealed disgust.
- People remaining in the organization are demoralized.
- The organization's return on investment (ROI) and/or return on assets (ROA) drops because of funding spent on process improvement increases without a measurable business return.
- The process improvement initiative experiences multiple false starts.

Years ago, I witnessed one of the worst cases of slash-and-burn process improvement. I led a CMM appraisal (CBA IPI)[5] on a medium-size unit (about 100 people) of a very large outsourcing company. After we delivered the appraisal findings, I was approached by few people who had been with the organization more than eight years and had been employees of the company that had the contract prior to the outsourcing. These "old timers" pulled me aside and wanted to tell me something important; I was listening.

What they said floored me! They said that, after eight years spent in process improvement, they were happy that their organization had finally achieved the maturity level it had attained before the outsourcing. Noticing my jaw hitting the floor, they filled me in: "Oh yeah, we had defined processes and used them to manage and control our work, but when [company name withheld] took us over, they told us to throw away all our processes and start doing things the [company name withheld] way." Overlook, for a moment if you can, the human toll — the senseless demoralization of a skilled workforce. Think of the tragic waste of money this particular slash-and-burn CMM effort represents. Name Withheld is a publicly traded company. Wouldn't you be even mildly concerned if you were a stockholder?

Fortunately, there is a better way to approach systems process improvement that is not only kinder and gentler, but also more cost-effective and efficient. It is a natural approach and it's partially described in the following section and then more fully addressed in Chapter 4 — Process Improvement Strategies that Work.

The rest of this chapter describes some of methods you can use to search for and recognize native practices in an organization that can be leveraged when implementing CMMI-based procedures. For more detailed information on cultivating and growing the organization's native processes and practices, go to Chapter 4 — Process Improvement Strategies that Work and read the sections titled, "Process Improvement: Good, Fast, or Cheap," and "Natural Process Improvement through Weeding and Nurturing."

Why People Slash-and-Burn

Who really knows, but based on a sampling of the characters I've watched employ this approach, here are some reasonable guesses at root causes:

- *They don't know any better.* This is the easiest excuse to forgive because in this case the pushy process people aren't malicious, just ignorant. They've never been exposed to any other type of approach to technological or cultural change, so they just go with what they know. If they're not wedded to the status quo and if they can learn something new without feeling threatened, there's a good chance of getting them to change.

- *They are control freaks.* Without getting too Freudian, you know the type. These are the people who insist on everything in the workplace being done a certain way — their way. Never mind that most people in this category live a personal life outside of work that is screwed up beyond repair. Sometimes, the more dysfunctional they are outside of work, the harder they try to beat others into compliance with rigid rules in the workplace. If their perceived control is threatened, they can be vindictive.
- *They are afraid.* Plain and simple, if the culture of the organization is one of command-and-control in which people do what the boss says "or else," don't expect to find a lot of creative thinking in play. It doesn't matter if the procedures are not useful or even disabling; the rules are the rules and you will follow them. A special category of this character commonly found in this environment is the "fear-biter," someone who attacks when pushed into a corner and is suffering maximum fear.
- *It's the culture.* There are cultures in which the slash-and-burn approach to process improvement seems perfectly normal and, in fact, aligns very well with the prevailing culture. For example, you'll often find the symptoms of the slash-and-burn approach in organizations in which a high percentage of the managers are former military. They are accustomed to giving and taking orders and following and having them followed without question. In such corporate cultures, it doesn't particularly matter if the orders make sense or even make money, they are followed. Questioning the orders and challenging the status quo, both of which are healthy to organizational growth and adaptation, are behaviors which are not appreciated and are usually punished by the management. Fear is the unspoken, unwritten motivator. A variation of this culture exists in other parts of the world in which the driving goal is achieving a CMMI maturity level as fast as possible at all cost.

CMMI'S PLACE IN THE PROCESS IMPROVEMENT UNIVERSE

The publication of CMM (for Software) in 1995 spawned the emergence of an entire industry. Within a few years, hundreds of organizations and many thousands of people worldwide were spending millions of dollars to improve their software processes using CMM's guidelines. CMM-based conferences, tools, books, and consultants seemed to pop up overnight, everywhere, and simultaneously. The publication of CMMI added fuel to the fire, increasing both the supply and demand for expertise and assets devoted to improving software, systems engineering, acquisition, and management processes.

The world of CMMI process improvement has grown so big that if you spend most of your professional hours living inside of it, you can easily come to believe that CMMI and process improvement are one and the same. Just like people living in the 15th century who believed that suspiciously arced horizon about 20 miles out to sea was the edge of the world, I come across many people who believe that the world of systems process improvement is bounded and defined by the 700+ pages of CMMI. Both beliefs are observably inaccurate.

The boundaries of our natural universe continue to elude all of us and even the brightest astrophysicists. As a society, we got over Europe not being the whole world, comforting ourselves with the knowledge that at least Earth was a significant astronomical body. That delusion too was smashed, but that was okay because at least we were the occupants of a galaxy that surely was a major player in the universe. A recent article in *Scientific American* explains how it is mathematically and physically probable — not possible, probable! — that there are many parallel universes, none lesser than our own.[6] This kind of information can really bruise a person's ego and sense of self-importance!

And so it is with process improvement. People want to believe that CMMI *is* systems process improvement in its entirety. We want to believe this, because then systems process improvement is bounded, known, and comfortable. Oh sure, the heretics among us will occasionally give a gratuitous nod of acknowledgment to those aliens residing on Planet ISO, or those in the IEEE nebulae or the Malcolm Baldrige asteroid belt, or on that old, dead world, TQM. But CMMI and its close twin moons — TSP[SM] and PSP[SM] — is, for all practical purposes, the process improvement universe, right?

Wrong. Neither CMM nor CMMI created systems process improvement no more than the Earth created the universe. Just the opposite occurred: process improvement created CMM and CMMI. If you step back for a moment, pause, take a deep breath, stop listening to the dogma and hyperbole that flows freely in this industry, and look close and hard at organizations, you'll find naturally evolving business and process improvements organically springing up all over the place. Sometimes these blossoming improvements can be correlated with process areas or practices in CMMI. Sometimes they spring up almost despite the destructive path of a misguided CMMI bulldozer driven by a slash-and-burn operator at the controls.

INSIDE MATURITY LEVEL 1

Xerox is the kind of company where it's easy to wake up one day and find that you've been working there for 10 or 20 years. However, unlike many of my peers at Xerox, I spent much of my professional life with the company, but not all of it. Prior to working for Xerox, I worked for Grumman, Lockheed, Unisys, The Travelers, and for myself. I had experiences with which I could compare and evaluate the Xerox experience. (My personal satisfaction with Xerox often stemmed from my knowledge of the horrors that lie just outside the protective confines of the big, red X.)

In Xerox, as in other organizations, there are EBPs that either evolved naturally or were planted and cultivated in the environment by some quality-oriented initiative such as Total Quality Management (TQM) or the Malcolm Baldrige Award, which Xerox won in 1989. In many instances, these EBPs become so entrenched in the environment that they become, over time, as essential but as unnoticeable as the air we breathe. In CMMI terms, these EBPs become institutionalized.

If you look at a forest from one perspective, you might only see its tall, strong trees. They're all slightly different and sometimes internally competing for resources. You can view an organization the same way. The trees are the business units, departments, divisions, or product lines. Or, viewing the organization through a CMMI lens, you might see requirements being managed, projects being planned and managed, things being placed under configuration management (CM), and so on.

•In both the forest and the organization, what you don't see from a more natural perspective are those things that bind all the components of the system together. In the ecosystem view of the forest, a studied look reveals the moss on the same side of all the trees, the movement and transfer of water and air, the recycling of the organic matter that makes up the forest floor for nutrients for the trees, the seeding of new growth and renewal, often through the death of a tree or a part of the total system. (For greater insight into applying a systems view or "systems thinking" to process improvement, read Chapter 4 — Process Improvement Strategies that Work.)

In a systems view of an organization, you will find the EBPs that integrate and hold together the boxes on the organization chart or the systems development processes. The forest is not just a collection of trees and a business is not just a collection of departments or processes based on CMMI.

So, what's inside a CMMI Level 1 organization and what are these EBPs? The answer is astonishingly simple. Most of the EBPs are activities or structures that you will find in just about any business organization in the world, whether or not it is engaged in CMMI-based process improvement.

I haven't found all of them yet, but here are the EBPs that play a significant role in CMMI-based process improvement:

- Writing and documentation
- Previous or concurrent quality or improvement initiatives
- Project management discipline
- Organizational standards
- Training programs
- Meeting management
- Measures and measurement activities

In CMMI, SEI made a significant improvement with the Continuous Representation by categorizing the process areas into logical groupings: engineering, project management, process management, and support. You could almost consider the Continuous Representation categories as "meta process areas."

Each EBP is yet a different kind of meta process area. Instead of grouping similar CMMI process areas, they group activities which act as common threads running through all the process areas. The common threads of the EBPs are interwoven through the CMMI process areas and can be used to weave together an integrated approach to using an integrated model for systems process improvement. (See Figure 1.2.)

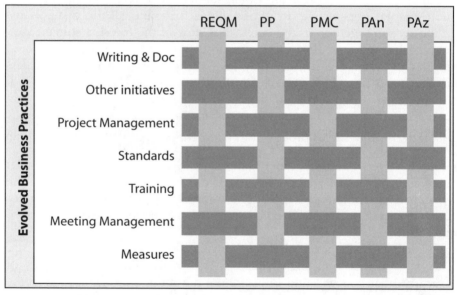

Figure 1.2   EBPs Interleaved with CMMI Process Areas

Let's take a closer look inside each of the EBPs and see how you can use them to improve your organization's process capability or maturity level using that which already exists in the organization.

Writing and Documentation

The following is a random list of activities or things you might witness in a typical system engineering organization:

- E-mail
- Project plans
- Database creation
- Software design document
- System requirements updates
- Employee performance reviews
- Training slides
- Contracts
- Written status reports
- Processes and procedures
- Template or form development
- Meeting agendas
- Program reviews

What do all of these things have in common? They certainly don't all fit neatly into one CMMI process area, generic goal, or even one process category. Here's the common denominator: they all involve writing or the existence of some form of document. (Note: A document does not have to be on a physical sheet of paper.)

Well okay, so why is writing and the ability to create documents so important to a CMMI effort? Two reasons: (1) most of the effort and cost associated with a CMMI process improvement project will be spent writing, editing, or revising some kind of document and (2) the availability of this skill is hugely overestimated in most organizations.

Let's first look at just how pervasive the work of writing or creating and maintaining documents is in the American workplace. I conducted a survey of 80 people chosen more or less at random. Their roles in their respective organizations represent most of those you would find in a typical systems organization. The roles surveyed include software, hardware, and system engineers, project and program managers, facilities and logistics support personnel, purchasers and contract personnel, process improvement managers/leads, senior managers, executives. The survey asked the participants to estimate the approximate percentage of their work time they spent in each of the following activity categories. Each category was defined and included examples of activities that fit into that category.

- *Thinking:* Formulating or manipulating ideas or thoughts entirely in your brain, but not performing any physical work or producing physical outputs.
- *Writing:* Includes revising, editing, formatting, or modifying any type of document that consists primarily of words.
- *Talking:* Speaking to others and listening to others speak but not performing any physical work or producing physical outputs.
- *Work other than writing:* This category includes activities such as writing software, building or assembling hardware, performing tests, monitoring network activity, etc.

The survey participants were given other guidance. For example, writing and talking also involve thinking but neither thinking nor talking necessarily includes writing. So, if thinking or talking also concurrently involved creating or changing any type of document, the participants were instructed to count the time in the writing category.

If asked if writing was their job, most of these people replied, "no." Yet people's responses to the survey suggest that writing is a big part of their work. The amount of effort spent creating or changing documents is likely to be even greater in process improvement work. No matter which phase your CMMI process improvement effort is in, people are writing. In a baseline appraisal, people are writing observations and findings. In process improvement planning, people are writing process improvement plans and subplans. After planning, people are creating or modifying policies, processes, procedures, and process implementation work products. In implementation, they're revising the process assets, creating training materials, sending mail notes to systems project personnel, and writing status reports. Come appraisal time again, they're writing an appraisal plan and then documenting observations and findings. Get the picture?

Now we get to the number two reason why writing and documentation are so important: these skills are usually lacking in organizations. Maybe the problem is that writing is either not viewed as a learned skill or that we just assume that everyone can write; we take it for granted. Ironically, we all tend to agree that writing software is a learned skill, that designing a database architecture is a learned skill, and even that managing people well is a learned skill. Yet we don't seem to give very much weight to the skill that is fundamentally critical to doing all of these and other jobs — written communication. After all, anyone with a high school education can write, right? No.

If you're in charge of your organization's process improvement effort and you're really lucky, you'll get to select the people who make up your process focus group, the SEPG or whatever you call it. You'll pick the best and the brightest people who represent and can speak for their functional areas: engineering, design, project management, quality assurance, configuration management, senior management, contracting, security, etc. You'll bring these really bright people together and the first thing you'll ask them to do is something they do not do well — write. You have practically set these bright people up to fail and, worse yet, look or feel incompetent.

This is exactly the scenario in one of the organizations with which I've worked. They managed to pull together some of the best people in the organization to form the enterprise-level process focus group. Based on a plan, these people started developing a really good, comprehensive yet tailorable standard set of processes. Prior to piloting their processes, they convened several times to peer review their own work. As indicated by the defect logs that were generated from these peer reviews, the vast majority (about 70 percent) of the defects and the vast majority of the rework and revision effort was expended on defects that could have been prevented if they had previously adopted some simple documentation standards for the process assets. Yet who knew or thought to recruit a technical writer or documentation specialist for the SEPG? I know now, but I didn't know then. Now you'll know.

Finding and Leveraging the EBP

There are two simple things you can do to reduce or eliminate the pain and cost of rework related to writing and documentation:

Find and recruit writing and documentation skills for your process focus group. Although technical writing or technical publication departments have become all but extinct, you can still find pockets of competence in this area in most organizations. Get help from such people for all of your CMMI work and get it early.

Many organizations have some documentation standards although — again — such standards might be a useless relic if no one abides by them. If you can't find any homegrown documentation standards, go elsewhere and get them for free. Once you have format standards — you know, headers, footers, type face, headings, margins, all that stuff — don't stop there! What makes documents, especially processes and procedures, either work or not work for people are their contents.

So, before you get highly paid people to sit around and fret about type sizes and acronym use, put their minds to something really useful like determining the content standards (what goes in the various process-related documents), the logical sequencing of sections, the applicability and scope of policies versus processes versus procedures, the use of graphics or illustrations, etc. You need to ask and answer questions like:

- What makes a process versus what makes a procedure?
- What goes in a policy?
- How does a form differ from a template?

And you need to write down those standards. (To address these specific questions, read Chapter 5 — Five Critical Factors in Successful Process Definition.)

Previous or Concurrent Quality or Improvement Initiatives

One of the more common features of the American corporate landscape is the continual and often simultaneous implementation of multiple process, product, quality, and strategic initiatives. Some companies and organizations will jump on almost any quality or improvement initiative that appears to be the trend or what the competition's doing. Some companies even take on multiple initiatives simultaneously, with corporate staffers sitting around dreaming up glossy materials to convince the workforce that this new thing will be really good for them and the company. The staffers in HQ can often have offices just a few feet distant from each other, yet it doesn't seem to occur to them to talk to their peers to find out if their separate initiatives can be somewhat integrated to save the company time, money, and employee aggravation. The observed approach appears to be, "No, let's just leave the simultaneous implementation of our multiple initiatives up others. We're the idea people."

Modern managers — in both the government and private sectors — are constantly tasked with having to integrate quality and improvement concepts and practices into their work, while still producing high quality deliverables faster and at lower cost. In the words of one parable, the modern leader is expected to "sharpen the saw" while simultaneously continuing to "cut the trees" faster and better. Today's leaders are inundated with initiatives. Six Sigma, Lean Sigma, Critical Chain, CMMI, ISO, Balanced Score Card, enterprise resource planning (ERP) and customer relationship management (CRM) are just a few of the larger initiatives on today's business landscape; tomorrow there will surely be more.

As this book is being written, Natural SPI is working with a government organization in this exact situation. The senior manager has been told by his bosses to implement the theory of constraints as described in Critical

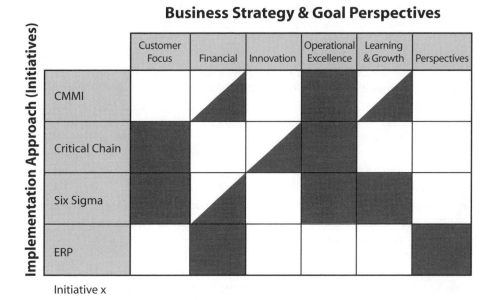

Business Strategy & Goal Perspectives

Implementation Approach (Initiatives)

	Customer Focus	Financial	Innovation	Operational Excellence	Learning & Growth	Perspectives
CMMI						
Critical Chain						
Six Sigma						
ERP						

Initiative x

Figure 1.3   Mapping Multiple and Quality Improvement Initiatives to Business Goals

Chain,[7] but also to incorporate CMMI practices, Personal Software Process[SM] (PSP) practices, and Team Software Process[SM] (TSP) practices into their system delivery operations. To start helping this client, we are building an array that will help him sort out which aspects of the multiple initiatives will help the organization achieve its goals. The first layer of understanding will look something like the illustration in Figure 1.3.

In the array in Figure 1.3, many of the typical business perspectives or areas of focus are shown across the top. These are the categories for grouping business goals that support the enterprise's strategy. The rows represent some current quality, process, or productivity improvement initiatives (Implementation Approaches). The partially shaded, fully shaded, or not shaded cell at the intersection of a perspective and an initiative represents the relative amount of benefit that perspective can reap from implementation of the initiative.

Through diligent research and dialog with experts, we are able to help this client to at least have a goal-level perspective of which parts of the different initiatives his organization should start looking into. Of course, there are more layers beneath this initial cut, but it does offer the advantage of saving the client the effort of having to read and internalize several thousand pages of models and academic books.

During my nine-year tenure at Xerox, the lower levels of the workforce, which included me, had to figure out how to execute Leadership Through Quality (LTQ), Software CMM-based process improvement, Integrated Supply Chain, and Time-To-Market (TTM). Moreover, there was constant pressure to integrate the principles and practices of these initiatives into our work, while continuing to develop and deliver products and while worrying constantly about the next layoff. In the Application Services Division (ASD) of CSC, we were trying to force fit CMM to everything the division did from common off the shelf (COTS) integration to pure help-desk functions, while implementing the strangest form of a balanced scorecard I've ever seen, while the division president flew around and made everyone read *Who Moved My Cheese?*[8] And the executives and senior managers of these enterprises were constantly wondering why we couldn't get better products out faster or keep their customers happy. We were literally "improving" ourselves into oblivion and you don't have to be a Harvard Business School grad to figure that out.

Corporate quality and process improvement initiatives have been around for a long time, at a minimum dating back some 20 to 30 years to the work of Dr. W. Edwards Deming and Joseph M. Juran. Most notably, three modern-day programs or initiatives require work that is often closely related to CMMI-based improvements: ISO 9001:2000,[9] the Malcolm Baldrige Quality Award,[10] and Six Sigma.[11] The synergy between ISO 9001 and CMMI and Six Sigma and CMMI are discussed in this section.

Before initiating a CMMI-based process improvement project, ask around and try to identify other quality or improvement initiatives that are either currently in progress or have been undertaken in the past. If ISO, the Malcolm Baldrige Award, or Six Sigma is in the organization's present goals or recent history, you are bound to find management and engineering practices already existing that will map well to CMMI. Leverage what these other programs have already accomplished. Don't undo what they have built only to spend time and money rebuilding something similar.

The following sections will give you enough information on the synergy between these separate initiatives to conduct a more thorough investigation on your own. A more thorough comparison of CMM and ISO and the Malcolm Baldrige Award can be found in Michael O. Tingey's book, *Comparing ISO 9000, Malcolm Baldrige and the SEI CMM for Software.*[12]

At any rate, the main idea is to spend a little time investigating the organization's history in improvement efforts. You can sometimes find a wealth of tribal knowledge and engineering practices that you can leverage.

ISO 9001:2000 and CMMI

Natural SPI, my consulting firm, enjoyed the wonderful opportunity of working with a very large-scale CMMI implementation on an Air Force program called J-Tech. J-Tech was essentially a range interoperability contract and involved implementing CMMI-based systems engineering improvements at four USAF test and training ranges. The contract called for the prime contractor — JT3, LLC — to achieve a CMMI maturity level and to be compliant with ISO 9001:2000.

From the outset of the contract, one of the most common questions people asked the JT3 CMMI Program team went something like this:

> I'm already doing this ISO stuff, so how does CMMI fit in? And are you going to make me do duplicate work to satisfy both models?

Usually some time after the respondent overcame the initial urge to snap back, "oh quit your whining," there came a more thoughtful, constructive answer. The better answer was:

> The CMMI model and ISO standards correlate and complement each other much more than they clash. The JT3 CMMI Program Team understands that and, with your help, will make sure you don't become a casualty of a models war.

Before people get all riled up about the differences between CMMI and ISO 9001:2000, get out in front of them by pointing out what they have in common. First and foremost, both of these things are abstractions of reality; CMMI is a model for operational excellence and ISO 9001:2000 is a standard for quality systems. As abstractions, both models/standards (let's collectively call them a "modard") identify practices that are commonly found in successful systems and service delivery organizations. They define what people do or what occurs in organizations that demonstrate measurable operational excellence. Neither modard explicitly tells people how to go about implementing the practices they define. In today's organizations, the job of determining how people will improve the way they work using the modard requires people to engage their brains; apply some intelligent interpretation to defining how the modard fits in the environment; and figure out ingenious ways to implement practices that satisfy ISO, CMMI, and actually help people do their jobs more effectively and efficiently.

There are more similarities. A review of mappings between ISO 9001:2000 and the CMMI from three sources indicates that cross-correlation of ISO clauses to one or more CMMI practices ranges from 60 to 90 percent.[13-15] That means that at the concept level there is overlap between the goals or intent of CMMI and ISO. This in turn means that if there's poor implementation of the modard in the organization, you could end up doing similar and redundant tasks or activities trying to satisfy both ISO and CMMI. So, how do you avoid creating a giant pile of documents that make you go off and perform two different things to achieve the same goal?

In the case of JT3's CMMI Program, the ISO effort had gotten a head start on the CMMI work. The people responsible for ISO-based implementation had developed numerous "practices," which were policy and procedural documents. Having these practices in place reduced the amount of work for the JT3's CMMI process focus group (called JPEG for JT3 Process Engineering Group). In many cases, all JPEG had to do was map the existing practices to CMMI or make incremental revisions to the practices to include systems engineering work and activities. Making minor revisions is always easier and more cost-effective than defining a process from scratch. The leadership of JPEG was also smart enough to include in the JPEG membership people who worked in the organization that lead the ISO initiative. These individuals' participation in both the CMMI and ISO realms played a key role in minimizing overlap and duplicate effort.

Are there some differences between CMMI and ISO 9001:2000? Sure, nothing's perfect. For starters, ISO 9001:2000 is 32 pages in length; CMMI-SE/SW/IPPD/SS, Version 1.1 is over 700 pages in length. Ostensibly, there might be a difference in the level of detail provided in these two documents. ISO 9001:2000 defines requirements for a quality system whereas CMMI provides implementation guidelines (e.g., typical outputs from a process) for each practice in each process area. The other major difference is that ISO 9001:2000 defines a minimum standard for organizations to achieve, whereas in the CMMI, different levels of operational excellence are defined via either process capability or organizational maturity levels (Continuous versus Staged Representations).

And, this story could go on and on and bore you to death with academic discussions about the similarities and differences between ISO and CMMI. But that's all it is, academic discussion. What should matter to you is that, in the end, you're not so buried in processes and procedures originating from the modards that you can't do your job and satisfy your customer. Find the synergy. Build the links and mappings. Don't duplicate effort.

Six Sigma and CMMI

This section will not teach you about Six Sigma, what it is, and how people use it. There are ample references on that topic. This section describes some of the relationships between Six Sigma and CMMI and how your organization might leverage those relationships if both these initiatives are underway. Again, when an organization is dealing with multiple quality or improvement initiatives, the goal is efficiency through limiting duplication of effort.

In a nutshell, Six Sigma is a management philosophy for improving organizational performance by reducing or removing variation.[44] The general business benefits touted by Six Sigma are greater process performance predictability, less waste and rework (lower cost), products and services that perform better and last longer, and happier customers. Hmm, those first two benefits — greater process predictability and reduced waste and rework — sound familiar, don't they?

Six Sigma is implemented via a path known as Define, Measure, Improve, Analyze, and Control (DMIAC). Within the CMMI process areas, there are practices which can be used to support the implementation of DMIAC. For example, the CMMI's Measurement and Analysis (MA) practices can be used to provide information in the Six Sigma Measure phase of DMAIC and CMMI's PMC, IPM, and QPM process areas can be leveraged in the Control phase of Six Sigma's DMAIC.

There are other comparisons between CMMI and Six Sigma. In general, CMMI identifies what activities are expected in a process. Six Sigma helps make those activities more effective and efficient. So, for example, an organization will use CMMI to make sure it is implementing activities such as estimating and risk management to improve overall project planning. Six Sigma can be used to continually improve the accuracy of estimating and risk management. CMMI-based process improvement efforts often have little impact on business results such as fewer defects, lower cost, and higher efficiency, whereas Six Sigma activities are totally driven to yield measurable business results in these areas. Six Sigma can provide specific tools for implementing the CMMI practices in DAR, CAR, and QPM. Six Sigma is weak in terms of establishing an improvement infrastructure whereas this is one of CMMI's strengths through OPF, OPD, OT, and the Generic Practices (GPs).

Project Management Discipline

Many CMMI practices — even those defined in the Process Management, Engineering, and Support categories — involve basic project management practices. Much of the "what to do" and "how to do it" of project and program management is also defined in the Project Management Body of Knowledge (PMBOK®),[16] an extensive repository of project management information established by the Project Management Institute (PMI®). PMBOK is used by thousands of project managers, many of whom are also involved in the planning and implementation of CMM or CMMI-based process improvement. The challenge for project managers is working with PMBOK and CMMI without duplicating effort to accomplish common goals and tasks.

This section gives you an overview of the synergy for reuse between CMMI and PMBOK. It discusses a case study in which both CMMI and PMBOK were jointly used to define an effective enterprisewide project/program management process. Specifically, what you can learn from this section is practical, actionable information on three main aspects of CMMI-PMBOK synergy:[17]

- A project management view for process improvement
- CMMI and PMBOK similarities and differences
- CMMI and PMBOK strengths and weaknesses

Project Management for Process Improvement

There have been times when I have asked someone how their process improvement "project" is going, and the response has been along the lines of, "what do you mean, 'project'?" or "it's not a project." So then I ask a few more questions like:

> Did your CMMI effort have a start date?
> *Answer: Yep.*
> Does your CMMI effort have an end date?
> *Answer: Yep.*
> Does your CMMI effort have a schedule, estimated and allocated resources, and will it deliver something?
> *Yep, yep, yep.*
> Are there risks to achieving your goals?
> *Yep.*

So, how can your CMMI effort be anything but a project? And if you accept that it is a project, why would you not use proven project management methods, practices, and techniques to manage it?

Just because a process improvement project delivers a "process system" and doesn't deliver software or software-intensive systems, there's no reason you can't use the basic concepts and practices of good project management. When you build and deliver a process system, you're delivering something internal to the organization that will help people do their jobs more effectively and efficiently, just as e-mail and voice mail systems helped people communicate better in the workplace. Both the applicable CMMI practices and PMBOK can help people with process responsibilities effectively plan and manage their process improvement project. How to better plan and manage your process improvement project is described in detail in Chapter 3 — Managing the Process Improvement Project.

CMMI and PMBOK Similarities and Differences

There are both similarities and differences between PMBOK and CMMI. The primary difference is that CMMI contains practices that provide guidelines applicable to almost the whole spectrum of system engineering and management activities whereas PMBOK's focus is on project management.

The similarities between PMBOK and CMMI are strong in the process areas in the Project Management category of the Continuous Representation, particularly Project Planning (PP), Project Monitoring and Control (PMC), and Risk Management (RSKM), with PMBOK being more rich in information on risk management than CMMI.

In the JT3 process improvement program, JPEG established process improvement teams, with one of the teams devoted to defining processes and work products for all the PAs in the Project Management category. This intuitively makes a lot of sense because of the strong inherent interrelationships between planning a project and then executing the plans. The Project Management (PM) team was fortunate enough to have two PMI-certified Project Management Professionals (PMP®) as members.

The team ultimately designed four major project management processes and associated implementation work products structured along the lines of PMBOK's four PM phases: (1) Project Initialization, (2) Project Planning, (3) Project Execution, and (4) Project Closure. The architecture for the project management processes was defined in a sophisticated Excel spreadsheet that correlated many process components including the process task or subtask, primary responsibility (by role) for performing the task, entry criteria and inputs, process implementation work products, and exit criteria and outputs. For each project management task or subtask, the table also identified the correlating CMMI practice and PMBOK practice. Table 1.1 shows an outline of the JT3 PM process and the CMMI and PMBOK relationships.

CMMI and PMBOK Strengths and Weaknesses

Both CMMI and PMBOK have their respective strengths and shortcomings. The main differences are described in the following subsections.

Table 1.1 Case Study: JT3 CMMI-PMBOK Relationship

Implementation Process	CMMI Practices	PMBOK Practices
Phase 1: Project Origination		
Originate/start project		5.1 Project Initialization
Initially scope and authorize project	PP SP 1.1, SP 1.2	5.2 Scope Planning 5.3 Scope Definition
Phase 2: Project Planning		
Develop Project Management Plan(s)	PP SP 2.7	4.1 Project Plan Development
Plan project life cycle	PP SP 1.3	2.1 Project Phases and Life Cycle
Plan project process (tailored JEEP)	IPM SP 1.1 GP 2.2 for IPM	3.4 Customizing Process Interactions
Estimate project effort and cost	PP SP 1.4	6.1 Activity Definition 6.2 Activity Sequencing 6.3 Activity Duration Estimating 7.1 Resource Planning 7.2 Cost Estimating
Develop project schedule	PP SP 2.1	6.4 Schedule Development
Plan project resources	PP SP 2.4, SP 2.5 GP 2.3 and GP 2.4 for PP, PMC, RSKM, and SAM GP 2.4	9.1 Organizational Planning 9.2 Resource Planning
Plan project stakeholder involvement	PP SP 2.6 GP 2.7 for PP, PMC, SAM, RSKM	2.2 Project Stakeholders 10.1 Communication Planning 11.1 Risk Management Planning
Plan project risk management	PP SP 2.2 RSKM SP 1.1, SP 1.2, SP 1.3, SP 2.1, SP 2.2, SP 3.1	11.1 Risk Management Planning 11.2 Risk Identification 11.3 Risk Analysis — Quantitative 11.4 Risk Analysis — Qualitative 11.5 Risk Response Planning
Plan project monitoring, control, and reporting	GP 2.2 for PMC GP 2.8 for PP, PMC, RSKM GP 2.10 for PP, PMC, RSKM	10.1 Communication Planning

(continued)

Table 1.1   Case Study: JT3 CMMI-PMBOK Relationship (Continued)

Implementation Process	CMMI Practices	PMBOK Practices
Plan vendor or COTS acquistion	SAM	12.1 Procurement Planning 12.2 Solicitation Planning
Plan work product verification and requirements traceability	VER	
Plan work product validation/testing	VAL	
Plan project measurements	MA SP 1.2, SP 1.3, SP 1.4 GP 2.8, GP 3.2	
Plan project CM/DM	PP SP 2.3 GP 2.6	
Plan project QA	GP 2.9	8.1 Quality Planning
Review and approve project plans	PP SP 3.1, SP 3.2, SP 3.3 GP 2.2, GP 2.3, GP 2.4, GP 2.5	
Acquire resources and budget	PP SP 3.2 GP 2.3, GP 2.4	7.3 Cost Budgeting 9.2 Staff Acquisition
Phase 3: Project Execution		
Monitor and report project performance against plans	PMC SP 1.1, SP 1.2, SP 1.3, SP 1.4, SP 1.5, SP 1.6, SP 1.7 MA SP 1.2, SP 1.3, SP 1.4, SP 2.1, SP 2.2, SP 2.3 GP 2.7, GP 2.8, GP 2.9, GP 2.10, GP 3.2 REQM 1.5 VER VAL	5.2 Project Plan Execution 5.4 Scope Verification 5.5 Scope Change Control 10.2 Information Distribution 12.5 Contract Administration 8.2 Quality Assurance 10.3 Performance Reporting
Analyze project performance, risks, and issues, and take corrective action	PMC SP 2.1, SP 2.2, SP 2.3 MA SP 2.4 GP 2.7, GP 2.8, GP 2.9, GP 2.10 REQM 1.5 VER	10.3 Performance Reporting 4.3 Integrated Change Control 7.4 Cost Control 8.3 Quality Control 6.5 Schedule Control 11.6 Risk Monitoring and Control

(continued)

Table 1.1   Case Study: JT3 CMMI-PMBOK Relationship (Continued)

Implementation Process	CMMI Practices	PMBOK Practices
Phase 4: Project Closure		
Complete project		12.6 Contract Close-Out
Conduct project lessons learned		10.4 Administrative Closure
Close project		
Terminate project funding instruments		10.4 Administrative Closure
Archive project work products	GP 3.2	10.4 Administrative Closure
Archive project measurements	MA SP 1.2, SP 1.3, SP 2.1, SP 2.3 GP 3.2	10.4 Administrative Closure

Project Origination

Starting with the Project Planning (PP) process area, project management ala CMMI seems to be based on projects just mysteriously appearing out of thin air. CMMI does not say where projects come from or how they originate; presumably they simply exist and you can begin with planning. PMBOK does a much better job of addressing the origination of a project and how it can be initially scoped in such a way that the project begins as a proposal. If the proposal is funded, it can then be planned; if not, it (appropriately) dies.

Project Planning

Both CMMI and PMBOK do a good job of covering project planning activities and resulting work products. CMMI is more comprehensive in consciously addressing the selection of a life cycle and developing a project process that is a tailored version of the organization's standard process(es). PMBOK takes CMMI to school on risk management planning, but CMMI kicks PMBOK butt in planning engineering and support activities such as requirements development and management, validation, configuration management, and process quality assurance.

Project Closure

PMBOK beats CMMI in this area also. Although CMMI addresses planning system deliver and transition into the user or customer environment, it falls short of addressing the administrative closure of the project. In a purely CMMI world, a project never ends. (Hmmm, I wonder if that aspect of the model is really an abstraction of reality?) PMBOK makes sure that you close project accounting numbers or charge codes, that the customer has accepted the product/system, and that the project has collected lessons learned.

Institutionalization

The concept doesn't even exist in PMBOK, so don't bother looking. CMMI's generic goals and practices are there to ensure that project management success can be repeated, irrespective of the players involved.

Organizational Standards

One of quickest ways to alienate software and systems developers and project leads is to come into their organization and start using the phrase "formal standards" as in "we need formal standards to use in peer reviewing work products." Almost without exception, every software or system engineer I've worked with completely tuned out the process people when they started using this "S" word.

The reason? Software and systems engineers today still view themselves and their craft more as artists and art and not an engineering discipline. (This self-image is misguided of course, but let's go along with it for now.) They view standards as something that inhibits or stifles the creative process of developing code. At this junction, you have really only two options: (1) align your efforts to institutionalize standards with the prevailing culture and beliefs or (2) prepare to fail.

 To align your efforts to implement standards, first change *your* language. Instead of talking about standards, talk about "the way we normally do things around here." Then, to spread those standards across the organization (i.e., institutionalize them), get people to take pride in knowing that they do a certain thing very well. Help them be proud enough to want to "immortalize" their expertise in the form of a document, so that others can benefit from "the best way to do something." Viola! An organizational standard is born.

Prior to engaging in formal process improvement initiatives, many organizations have evolved over time a set of "standards." Such standards may take many different forms including the formatting of various types of documents, the way people answer their telephones, work start time, what can or cannot be shown on a business card, how code is checked in and out of a configuration management system, what can be displayed (or not) in cubicles and offices. Such standards or standard practices may often exist simply as an aspect of the culture that has been around for a long time. They may not be documented anywhere, but nevertheless, people know about them and follow them.

Those same systems engineers who will tell you to your face that they "don't need any stinking standards," will, if coaxed or approached in a nonthreatening way, quite proudly tell you that they've developed the "right" way to do something over the years and that everyone in the organization should do it their way. Your instinct will be to dismiss it as arrogance, but you and your organization will be much better off if you recognize this for the goldmine it is — you have powerful advocates of standards so long as you can avoid using that "S" word. These organic standards can serve you as a significant leverage point in your process improvement effort. Let's look at some examples.

Say your organization already has a default way status reports are formatted. By "default" I mean that if you were to look at a random sample of five to ten status reports, they would all have a similar look and feel, and contain similar content. Since these reports are the primary tools people use to communicate their progress and activities, this standard has saved you and the organization significant time and effort because you now don't have to figure out, or "invent," how to write a status report.

Now you have to arrange to deliver some training. You ask around and find out that there's a particular administrative aid, Bill, who takes particular delight in making all the logistical arrangements for meetings and training events. He loves to secure the room and the equipment, arrange for food, and he'll even send out reminder e-mail to participants. This work isn't written down and it's not even a part of his documented job description, but he's really good at it and that's why everyone turns to him to do this stuff. Knowing this, why would you conceive of arranging your training logistics yourself? Why would you not delegate this work to Bill? And wouldn't you invite Bill, who is obviously a relevant stakeholder in matters concerning training, to participate in defining processes for the Organizational Training (OT) process area?

Remember the last time you were new to an organization, like when you last changed jobs? Remember how, in the first few weeks, people around you did things, but you couldn't discern why or how? It took you some time, maybe months or even years, to see the patterns in the way people in the organization worked. The same disorientation may occur when, after working in an organization for 20 years, someone new asks you why and how a particular task gets done. You have been working within the organizational norms (read: standards) and practices for so long that they have become part of your subconscious — they are automatic. It is now difficult for you to consciously articulate to someone else who doesn't have the same experience.

Now here's the real leverage in existing organizational standards: as you observe people following standards (cultural norms, standards practices, recurring ways of doing things, etc.), write them down and start collecting them. It's natural to go into an organization with which you're not familiar and not see any visible (documented) standards and assume there are none. Do not make this assumption; it is almost always inaccurate. Quietly observe people doing things, ask questions, look for patterns of behavior, and then confirm the existence of a standard by asking several people. Write it down and get a few others to read it and give you feedback on whether or not you've captured the standard correctly.

Here's an example of how to do this. You're planning to deliver an orientation on configuration management to people in your organization. You send out e-mail announcements to everyone in the organization. People show up and give you positive feedback. A week later, your boss comes to you, and has a conversation that goes something like this:

BOSS: Good job on that CM orientation; there was a good turn-out.

YOU: Thanks.

BOSS: By the way, why wasn't our marketing rep there?

YOU: Uh, well, I didn't send him a notice since he's not really part of our organization.

BOSS: Well, okay. But in the future, make sure you copy him on any CM matters. He always has approval on things having to do with product configuration.

Okay, you just learned a standard related to stakeholder buy-in on a certain aspect of your process improvement work. Write it down. Now, take it one step further. Go do some investigating and ask some questions. Find out who else within and without the organization normally participates in what activities/decisions and what exactly their participation is (eg., review, approval, input). The document that results from your investigation — let's call it a "Stakeholder Involvement Matrix" — will be of great value to you in two areas. First, it serves as a documented standard for identifying who participates in what and what their level of participation is. Second, it serves as a valuable tool in defining the intergroup relationships and dependencies, which will be integral to your organization implementing the Integrated Teaming (IT) process area and go a long way to establishing Generic Practice 2.7 — stakeholder involvement.

Remember, organizational standards don't just one day fall from the sky. Standards come into existence when someone such as you takes the time and interest to document the subject matter expertise that is inherent in the organization and then gets people who are going to use the standards to agree that

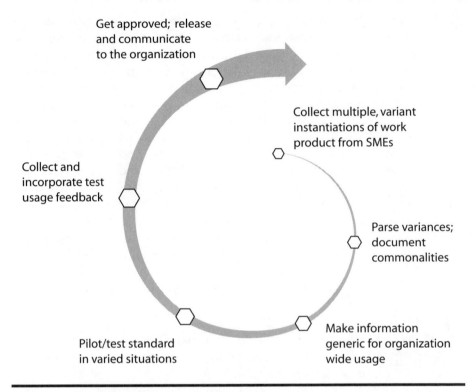

Figure 1.4 A Development Life Cycle for Organizational Standards

they represent the collective expertise and knowledge. Figure 1.4 illustrates the major steps involved in the evolution of a standard from its origin as an organic, assumed practice to a conscious, defined standard. These steps are described in detail following the illustration.

1. Talk to several people who frequently have to perform a similar task or develop a similar work product. A good example is project managers who frequently have to create a project plan for their projects. Gather multiple, variant examples of the work product from different projects. Document the similarities and differences. Talk to the project managers and get answers to questions on their usage of the project plans, such as:
   ■ What did you do differently and why?
   ■ What did you like about the project plan? What did you not like? What parts do you consider important? Which ones are not useful or don't add value to your job?
   ■ If you could have a template for doing the project plan in future projects, what would it look like? What would it contain?

2. Assimilate, analyze, and parse all the information you collect in Steps 1 and 2. Find the common "best practices" from all the instantiations of the work product and put them all into a pilot template. Also incorporate solutions from the answers to the questions asked in Step 2.
3. Remove any project-specific information (unless you identify it as "example" for instructional purposes), making the document generic for use in a wide variety of projects.
4. Get the project managers together and walk them through the project plan template. Tell them that you've tried to compile the very best of the work they've done in the past and that you've tried to remove problems from the software development plan (SDP). Tell them that you need the new template to be tested out on some projects to see if it works. Get senior managers to publicly encourage people and provide them with incentives for testing new work products and providing feedback to the people managing its development and deployment. Test the new work product (SDP template in this case). Provide coaching and mentoring to project mangers on its use and collect their feedback on usage.
5. Incorporate the feedback into a new version of the work products. Collect any compliments or endorsements on the work product from the users.
6. Get approval, perhaps from the SEPG, to make the new work product available to all users in the organization. When you announce the new item, make sure you advertise that it represents known best practices and that it has been successfully tested.

Training Programs

If you work or have ever worked in an organization larger than 100 people, I can guarantee you that you don't have to start from scratch in developing and implementing a training program (OT). Even my small consulting firm has an informal (defined but not documented) process for planning, obtaining, tracking, and measuring training required by our employees. As with standards, organizational education and training programs and infrastructure often lie just below the visible surface of the organization and you may have to poke around to find the structures that have naturally evolved. The following subsections give you some clues for where to look. Don't forget to take your pen and notepad and write down what you find. Those notes will later become the organization's defined process (GP 3.1) for OT.

Training Personnel, Tools, and Budget

Even the smallest organizations usually allocate some number of hours per year to employees for training (OT SP 1.4 and GP 2.3). You'll even frequently find homegrown or off-the-shelf training databases used for planning and tracking training (OT SP 1.3 and SP 2.2). Then go look for the organization's preferred vender list (or ask someone who does purchasing if there's not a documented list), and you may find the names of vendors from whom the organization has acquired training. Figure out how they got on the preferred vendor list and you'll have information for OT SP 2.3.

Meeting Management and CMMI

Meetings are a tool and, depending on how the tool is used, can be a productive use of people's time and achieve measurable results or they can be a colossal waste of time and resources. Meetings are also the place where many decisions are made.

One of the many reasons meetings can be unproductive is because different people have different needs for attending meetings and therefore have different success criteria for meeting outcomes. These different needs can result in people meeting at cross-purposes and collectively not achieving anything as a team.

Here is a subset of some of the reasons people meet:

- Make decisions
- Assign tasks and action items
- Get status on things; find out what's going on
- Resolve issues
- Plan work
- Talk with other people
- Be close to other people
- Get a sense of belonging to a group
- Exert influence on others
- Entertain others
- Be entertained by others
- Argue or debate with others (as a form of entertainment)
- Establish an alibi for not doing something else

So, if you hold or attend a meeting to make decisions and assign action items, but you're outnumbered by others with different purposes for the meeting, your success criteria has a good chance of not being met.

Productive meetings play a large role in implementing CMMI-based processes. Table 1.2 provides a short list of the typical types of meetings

Table 1.2 Typical Meetings and Related CMMI Process Areas

Typical Meeting	Correlating CMMI Practices
Requirements JAD sessions or meetings with customers to clarify work requests or requirements	REQM SP 1.1, SP 1.2 RD SP 1.1, SP 1.2 RD GP 2.7
Project team meetings	PMC SP 1.6, SP 1.7, SP 2.1, SP 3.2 PMC GP 2.7, GP 2.8
Senior staff meetings	GP 2.10 for all PAs
Project planning or work estimating meeting	PP SP 1.4, SP 2.2, SP 2.4, SP 2.6, SP 3.3
Meeting to review project plans	PP SP 3.1, SP 3.2, SP 3.3 PP GP 2.7, GP 2.8
Budget, resource or staffing meetings	PP SP 3.2 PMC SP GP 2.3 for all practices
Project reviews	PMC SP 1.5, SP 1.6, SP 1.7, SP 3.1, SP 3.2, SP 3.3 GP 2.7 for all practices GP 2.8 for all practices GP 2.10 for all practices

you can observe in almost any organization and CMMI practices that can correlate to the meetings depending on the content and agenda. Physical inputs and outputs associated with these meetings, such as agendas, minutes, and action items, can be used in an appraisal as direct and indirect evidence of the practices' implementation.

Where Everything Can Happen; Where Nothing Can Happen

The paradox of meetings is that they have such great potential; they are the place and time where everything can happen. Yet they frequently don't live up to their potential and become the place and time where nothing happens. Your organization will inevitably conduct meetings related to CMMI or process improvement work, so you'll want to make meetings productive. And since it is likely that many of the decisions that affect the process improvement work will come from meetings — such as SEPG meetings, steering committee meetings — you need to make sure meetings become an environment where decisions can be made.

One of the more positive experiences I had with Xerox was its culture for meeting management, a piece of the company's LTQ program that won the company the Malcolm Baldrige Award in 1989. Xerox had meeting management down to a science long before the company ever began using CMM for process improvement.

Meeting management at Xerox consisted of a set of practices that over time became ingrained in the culture. These practices were designed to make meetings purposeful, useful, and efficient, as opposed to the complete waste they are in many organizations.

A close friend of mine experienced meeting management culture shock when she left Xerox to work for another company. Within a very short time after joining the new company, she became involved in a series of Joint Application Development (JAD) sessions to define and approve some system requirements. According to her, these JAD sessions were inefficient and people spent a great deal of time to achieve minimal results. The reasons: they couldn't stick to an agenda (if there was one), no one kept the discussions focused, and agreements were difficult to reach because there was no decision process such as consensus.

Many of us living in the corporate environment loathe meetings, primarily because we find them less than useless. Yet we have no choice but to accept that meetings are a way of life, and we must live with them. As part of Xerox's LTQ program, meetings were transformed from barely tolerable events to gatherings that were productive and useful.

How many times have you sat in a meeting in which the first 15 to 30 minutes were spent trying to remember the decisions and actions from the last meeting? Now multiply that wasted time by the hundreds or thousands of meetings conducted across the company every week. The waste of time and money really starts to add up.

Xerox made meetings productive by institutionalizing two cardinal rules for meetings:

1. An agenda for the meeting had to be published and distributed to participants in advance of the meeting.
2. Meeting minutes, particularly decisions and action items, had to be recorded, and then distributed to participants after the meeting.

The first rule ensured that people didn't just get together and *then* figure out what to discuss. It was wonderfully self-regulating because people felt safe to not attend a meeting if there was no agenda, regardless of rank or status of the person calling the meeting. The second rule ensured that what was decided or assigned wouldn't have to be remembered and possibly repeated in a future meeting; it guaranteed outputs and follow-up work.

Structuring Meetings for Real Improvement (and CMMI Implementation)

With one client, we attended several meetings of the group responsible for planning, implementing, and governing CMMI-based process improvement.

There were no meeting agendas, minutes, or action items because the client felt there was no payback on creating such documents. There also was no defined decision-making process. Not surprisingly, they would find themselves discussing the same topics and issues for weeks on end, meeting after meeting, with no definitive resolutions or conclusions.

So one of the first process improvement actions we helped them plan was called "meeting management." You will not find a process area in CMMI called "meeting management," but we knew we had to make meetings work before we could expect to achieve any other improvements.

We got them to try out some simple Word templates for documenting the meeting agendas and minutes. Table 1.3 shows an example of a meeting agenda at the top and then an example of the resulting meeting minutes below.

At the time of the meeting, the agenda is easily converted into the meeting minutes by simply changing the heading of the last column from "Expected Results" to "Results."

To us (the consultants) it appeared to be simple and low effort to use these templates and, in fact, we knew this to be the case because we had used them in other organizations. But the client was still reluctant. So, we used a technique that has a high success rate, but which very few process consultants are willing to do: we, the consultants, rolled up our sleeves and did the work for them the first few times until they became "addicted" to the benefits. For several meetings, we produced the agendas and minutes. The client personnel quickly became accustomed to seeing and using these documents to structure their meetings. By the third instantiation, they had assigned a member of the group to take over the role of "scribe" to coordinate and publish the meeting agendas and to record and publish the minutes and action items. By the fifth meeting following the introduction of the templates, they were no longer rehashing old topics week after week and had gotten into a routine of efficiently processing issues. Meeting management had become institutionalized almost immediately.

Do all meetings need this level of structure? Of course not. There are types of meetings in which the only expected result is the free exchange of ideas, and there is no intention for their application, decisions, or for follow-up actions. But if you have a meeting from which you expect real results, not rambling; decisions, not dilly-dallying; and actions, not apathy; then put some structure around the meetings.

Meeting Rules and Guidelines: The Genesis of DAR

It's too bad that the CMMI Staged Representation has the Decision Analysis and Resolution (DAR) process area in Level 3, because it is one of those underappreciated PAs that would come in real handy right from the

Table 1.3   Sample Meeting Agenda/Sample Meeting Minutes

Systems Engineering Process Group Meeting Agenda for 6/16/03

Ref #	Topic	Leader	Type	Expected Results
1	Review minutes from last meeting and status action items	J. Abernathy	I, STA, D	Status and update action items as required. Expected Time: 15 minutes
2	Present and discuss as necessary NAVAIR Process Improvement Program **Preread**: NAVAIR SPI brief 20030604.ppt	Natural SPI	I	SEPG members to comprehend NAVAIR's overall CMMI process improvement policy, strategy, and structure Expected Time: 30 minutes
3	Review and concur to implement QA checklist for peer reviews Preread: QA Peer Review Checklist, V1	L. Nguyen	D, A	Collect final comments/changes to checklist and obtain SEPG consensus to add to PAL baseline and implement. Expected Time: 10 minutes

Agenda Types: I = Information sharing, D = Decision, A = Action, T = Training, STA = Progress/Status report, R = Risk, M = Measurements

Systems Engineering Process Group Meeting Minutes for 6/16/03

Ref #	Topic	Leader	Type	Results
1	Review minutes from last meeting and status action items	J. Abernathy	I, STA, D	Status and update action items as required. See PEG action item log.
2	Present and discuss as necessary NAVAIR Process Improvement Program **Preread**: NAVAIR SPI brief 20030604.ppt	Natural SPI	I	Presentation resulted in discussion and questions. **See action items 03-112 and 03-113.**
3	Review and concur to implement QA checklist for peer reviews Preread: QA Peer Review Checklist, V1	L. Nguyen	D, A	Checklist approved; Jeff to place in PAL and send out announcement.

Agenda Types: I = Information sharing, D = Decision, A = Action, T = Training, STA = Progress/Status report, R = Risk, M = Measurements

beginning of your organization's process improvement initiative. If I could go back and change one thing about how I've implemented process improvement in organizations, it would be to implement processes based on DAR a top priority and get this one nailed down before taking on any other improvements.

Why? Because everything you will do — from planning and conducting a baseline appraisal to developing a process improvement plan to executing that plan; everything — will require decisions to be made. Every action, every result, every work product, every component of every process, every word spoken, every hurt feeling, every success and every failure will be the result of a decision (or a decision to not make a decision). Business is — if nothing else — about making decisions which in turn drive actions. Doesn't it make sense to put some structure and predictability around how decisions are made in your organization?

It would be easy to read the DAR process area and say, "oh, we need to go build a decision process." Here's a news flash for you: people in your organization are already using their own processes to make decisions. These processes are most likely not documented and are likely executed at the subconscious level of thought. Still, you don't need to create a decision process; you need to go and explore people's minds and find out how they make their decisions and write it down. You need to ask questions before you start giving answers. Let's take another look at meetings and see what we can find in the way of decision-making processes.

People in organizations frequently made decisions but have no defined process, guidelines, or criteria for how decisions are made and every decision is made differently than the previous. This situation is the norm in most organizations.

Yet, with close observation (which requires you to suspend the belief that only you know "the right way to do it") you can see evolved decisions processes at work in meetings. Attend a few meetings and quietly take notes on observations such as:

- What kinds of decisions are made or topics discussed in meetings? Which topics are never discussed in meetings but are topics which obviously require decisions to come from someone?
- How many times did people look to a certain person to speak when a question came up about topic X?
- Did everyone have a chance to speak on a certain topic? By topics or classes of topics, who spoke and who did not?
- Who was in attendance when decisions were made? Which decisions were made without certain people in attendance?
- Does it look like certain people influence the decisions or votes of others in the room on certain topics? Who is influential and in which topics?
- Are decisions just assumed or does someone ever actually say something like, "okay, so what we decided was ..."?
- What kinds of decisions appear to be made but later are frequently reversed, ignored, or not understood?

After a while, you'll see some patterns emerge. No one ever argues with Sara on requirements; what she says goes. Jeff solicits comments from people on budgetary matters, but makes a decision regardless of dissenting opinions. People generally don't accept the marketing manager's view on system design. The group won't make decisions when more than two members are missing. On the average, two out of five decisions need to be readdressed later. From these patterns, you can begin characterizing the decision-making process with generalizations such as:

- Majority rule is the process for most decisions.
- Decisions involving funding or money are made by N, with input from others but not influenced by others.
- Changes to plans seem to require consensus.
- Low-level implementation decisions are left up to the individual engineers.
- Management commitments made to external individuals or organizations without prior discussion with the staff are usually ignored and not kept by the organization.

Once you have documented your characterizations (without naming names if possible), show them to the group you're trying to help. Let them know that you made observations that support the characterizations, how many observations, and on which dates. Try to get a general agreement from the group that your characterizations of their decision-making is close to accurate. If they agree it's accurate, find out which characterizations they feel good about and want to keep and which ones they don't like and want to change.

At this point, you have enough information to begin defining decision processes or guidelines for the group. You also have some baseline measures with which you can compare future measures (after implementing changes) to see if things improved and by how much. You don't need to invent a decision methodology; there are plenty of them already documented in books and online. The four decision-making methods used most frequently in business are:

1. *Autocratic:* One person or position makes certain decisions alone. She or he may or may not consider input from others in the decisions.
2. *Majority rule:* The group or team members vote and the decision is carried by 51 percent or greater of the votes.
3. *Consensus:* All members of the group must agree to support the decision. Support doesn't necessarily mean everyone likes the decision; it just means they're okay with abiding by it.

4. *Weighted voting:* Weighted voting is best suited to decisions in which there are multiple or tiered choices such as prioritizing goals. Based on the number of items to rank or prioritize, each member of the group is given *n* number of votes and can distribute their votes among the choices based on rules. Weighted voting should not be used for situations or questions requiring a simple yes or no decision.

In the Tactical Air Range Integration Facility (TARIF; China Lake, CA), we used a variation of the approach previously described. After observing several meetings of their Systems Engineering Group (SEG) which is responsible for, among other things, organizational process improvement, we discovered that they had naturally evolved a form of consensus and had developed an intuitive sense of quorum. Based on these observations and characterizations, we were able to help them define a decision-making process for the SEG, which was documented in the SEG's Charter and Processes document. The process has since been modified slightly, but its basic tenets are still functioning.

Measurements and Measures

Sometimes when I talk to people about measurements and measures, they act as if business, process, and product measures are something new, something introduced to the world by CMMI. Yet, once again, if you pop your head up out of CMMI and look around, you'll see that organizations have been finding ways to measure performance for quite some time. Think about it: What is a company's annual financial statement — required by law for publicly-traded companies — if it is not a collection of performance measures?

There are a number of sources of proven concepts and methods when it comes to process and product measures. The three sources I have found most useful in measurement work are Practical Software Measurement (PSM),[18] Goal-Question-Metric (GQM),[19] and Balanced Scorecard.[20] Balanced Scorecard, or just "Scorecard," is one of the more robust and provably worthy schools of thought that has caught on in American business today. It is a collection of processes, methods, and sometimes tools that organizations use to collect, report, and analyze performance measures in multiple areas of business to determine the extent to which the organization is achieving its goals, implementing its strategy, or fulfilling the vision. Like implementation of CMMI-based processes, there are as many different manifestations of Balanced Scorecard as there are organizations using it. There is substantial literature devoted to all three of these sources of measurement knowledge, which won't be repeated here. What you will learn here is how CMMI, GQM, and Balanced Scorecard can all fit together to give your CMMI effort increased visibility and credibility by tying it to the organization's business success.

Everything Starts with the Strategy and It Starts at the Top

Here's a mantra for you to remember and repeat. The more you play this mantra in your head, the more you begin to see natural, simple beauty of the idea. It is a cyclical path that can help people in an organization ensure that what they're doing always has meaning, purpose, and direction.

> Strategy and Vision beget Goals.
> Goals beget Action Plans.
> Action Plans beget Implementation.
> Implementation yields Measures.
> Measures indicate the accomplishment of Strategy and fulfillment of the Vision.

Here is something else you should realize: Balanced Scorecard gives organizations a way to communicate the strategy and goals and then measure performance against that strategy. CMMI gives you guidelines for establishing measurement programs (primarily through the MA, OPP, and CAR process areas), but also must start with the strategy from the top. GQM (and GQM-RX and GQM-Lite) is a tool for deriving measures based on the organization's goals which — no surprise here — come from the strategy.

So, if Balanced Scorecard starts with strategy and goals, and CMMI-style measurements start with strategy and goals, and GQM starts with strategy and goals, where do you think might be an important place to start your thoughts and plans about measurements?

Do not start your CMMI-oriented measurement work by forming a process action team that will then spend months defining a measurement or metrics process which, in turn, will force people to collect all kinds of measures that are never used by anyone. Start by talking to the leadership in your organization to learn about the overall business strategies and goals. If you cannot find any business strategies or goals, or no one knows if they exist, you've got much bigger problems than not having a measurement process.

But that's not a likely scenario; most businesses have something that can be called a strategy or things that can be called "goals," even if people cannot articulate very well what they are. Search through documents for these words: charter, objectives, targets, annual, growth, performance, metric, quad chart, and Scorecard. Finding any of these words can lead you to finding the organization's strategy or goals. Talk to people. Ask them what they're currently measuring and reporting, how they're doing it and why, and to whom do they report which data. I haven't walked into an enterprise in America yet in which no one ever had to report any kind of measures or performance information. Again, don't throw away or ignore what is already in place; acknowledge it, understand it, and build on it.

A Holistic Measurement Program

Figure 1.5 depicts the relationship between three sources of measurement-related methods: Balanced Scorecard, CMMI, and GQM.

As you can see in the illustration, the highest level construct that simultaneously holds together and drives all goals, plans, and actions is the organizational or enterprise strategy. Organizations often divide the strategy into different perspectives such as Customer Focus; Finance, which includes things like ROI, ROA; Innovation; Operational Excellence; and Learning and Growth. Within the perspectives are goals which, in essence, are specific aspects of the strategy and their achievements can be verified through measures or indicators. Balanced Scorecard provides a method

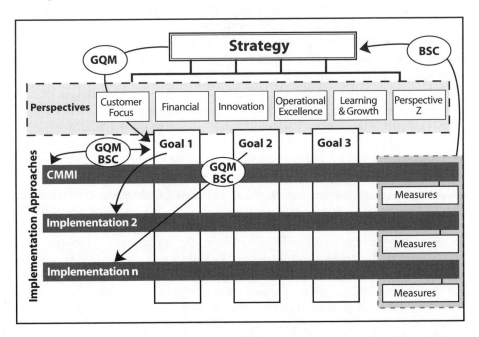

Figure 1.5   BSC, CMMI, and GQM Relationship within the Context of a Business

for driving and communicating the strategy down through all the hierarchies of the organization so that business goals, targets, and measures can be allocated to the most appropriate business units. A form of GQM is used to translate the organizational goals into action plans (see "GQM Lite: A Quick-Start Approach for Process Improvement" in Chapter 3 — Managing the Process Improvement Project).

Action plans for improving quality, process, or productivity can be assigned to different implementation approaches or initiatives (e.g., CMMI, ISO, Six Sigma), and different business units (e.g., R&D, marketing, manufacturing). The execution of the action plans within the various implementation approaches or initiatives can and should yield measures which, in aggregate, indicate the extent to which the goals are being met and the strategy is being accomplished. Scorecard and, perhaps, the organization's MA and OPP processes are used to define the collection, analysis, use, and reporting of the measures from their lowest level of collection to the executives and senior managers who established the strategy.

CMMI-based process improvement vis-à-vis the MA and OPP process areas is just one subgroup or category of process and product measures. Measures and indicators can and should also come from other implementation approaches or organizational units such as marketing, training, or human resources.

In JT3, JPEG made this concept real. Through much analysis, the group was able to establish and graphically depict the linkage between the lowest level process and product measures all the way up to the enterprise strategy. JPEG used GQM to derive the project level measures (53 in all) that would result from projects using the JT3 Enterprise Engineering Process (JEEP; a.k.a. "OSSP"). All 53 base and derived measures are collected and used by the projects. The local Engineering Process Groups (EPGs) at the four test and training ranges (equivalent to business units) and the JPEG use the 53 project level measures to derive only a handful of indicators, which are reported upward as supporting the JT3 contract performance indicators. The contract performance indicators are then mapped to broad statements in the JT3 statement of objectives, which is the enterprise's strategy. The concept of this linkage is illustrated in Figure 1.6.

As depicted, the relationships between the measures in each stratum of measures, the indicators, and the organization's objectives and goals can be one to one, one to many, many to one, or many to many.

The power and benefits of a tool such as the one built by JT3 in Figure 1.6 are significant and include:

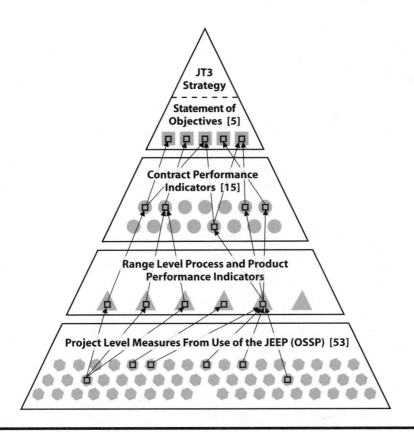

Figure 1.6   Case Study: JT3 OSSP Measures Linkage to Business Goals and Strategy

- It clearly answered all the questions related to the collection, analysis, and reporting of measures: what gets collected, by whom, when, and how? Which measures get reported up the hierarchy and to whom and why and which measures do not get reported upward?
- It gave the people involved in the process improvement work a strong sense of purpose and meaning for their work that had little to do with maturity levels.
- It showed how the CMMI-based process improvement work will benefit the entire enterprise.
- It gave everyone a tool for analysis. It became easy to determine the impact of not collecting certain low-level measures — "if we don't collect this, we cannot determine if Goal X is achieved."

DO'S AND DON'TS

Here's a summary list of things you should do and things you should not do when first starting out on your organization's process improvement journey using CMMI:

Do

- Assume that the organization in which you want to make improvement has already evolved some good management and engineering practices that can be incrementally improved using CMMI as a guide.
- Identify, acknowledge, and recognize the existing good management and engineering practices or process strengths.
- Identify any existing organizational standards that you can leverage.
- Look for evolved business practices in areas of the business that aren't specifically identified as process areas in CMMI.
- Find out what the organization's history (if any) for change and process improvement is.
- Base the documented processes and procedures first on what people are already doing and then begin the improvements.
- Have an open mind; learn as much as you can about the organization and its people before you try to make improvements or implement CMMI.

Don't

- Don't assume that the organization already has some good engineering or management practices or standards already in place. (Even if you think this, don't say it publicly.)
- Don't throw away or ignore evolved business practices that can be used as a springboard for CMMI-based improvement.
- Don't assume that just because people in an organization have been doing something or doing something a particular way that it still has business value. Legacy practices and standards can sometimes be a barrier to process improvement efforts.
- Don't take the attitude that the organization is in terrible shape and the only thing that will save it is you and CMMI — you will absolutely fail every time!

QUIZ: WHAT DID YOU LEARN? WHAT WILL YOU DO?

Now take the post-chapter quiz (Figure 1.7) and think about what you've learned and how some of your views toward CMMI-based process improvement have changed. Think about what you will do with the information you've learned (and how it makes you feel).

1. **An organization that has not been appraised at a CMMI maturity level:**
   a. Is in total chaos
   b. Has no processes
   c. Usually has evolved business practices that can be leveraged for CMMI-based improvement
   d. Is immature

2. **True or False:** The best way to implement process discipline is to write new procedures and make people use them.

3. **Process improvement requires which of the following skills:**
   a. Software and systems development
   b. Management
   c. Writing and documentation
   d. Configuration management
   e. All of the above

4. **True or False:** My organization can't follow what other organizations have done with the CMMI because we're different.

5. **The best goal for CMMI-based process improvement is:**
   a. Achieving a maturity level
   b. Achieving a process capability
   c. Creating new procedures
   d. Helping people become more effective and efficient in their work

Figure 1.7   Chapter 1: What Did You Learn? What Will You Do?

2

THE ROLE OF ROLES

I'm not asking you what you do, Dave. I'm asking you who
you are.

— **Jack Nicholson as Dr. Buddy Rydell in *Anger Management*[50]**

WHAT DO YOU THINK? WHAT DO YOU BELIEVE?

This chapter is about defining the roles of people in the organization and the roles of organizations which work together to deliver a system or service. As you will soon learn, it is very easy to make assumptions about roles and responsibilities in an organization. If left unchallenged — like comparing your assumptions with reality — the assumptions will soon become hard-held beliefs. Take the quiz in Figure 2.1 to see what kinds of beliefs you may have about organizational roles and then, after reading the chapter, take the quiz in Figure 2.3 to find out what you've learned.

THE MODEL AND THE REALITY

Both SW-CMM and CMMI provide either explicit or implicit guidance for establishing roles and responsibilities in the organization. In SW-CMM, there are Ability to Perform key practices that address establishing certain functions to perform processes and activities such as project management, change control, quality assurance, and training. In CMMI, there are no explicit practices that come right out and say, "define your roles and responsibilities," but there are several GPs that heavily depend on some level of established roles. GP 2.3 requires that resources be provided for performing the process, developing the work products, etc. A resource doesn't have to be a person or people in the abstract world of CMMI, but resources are almost always people in the world in which we work.

1. **True or False:** The words "roles," "responsibilities," "job," "title," and "position" pretty much all mean the same thing.

2. **The biggest problem my organization has in terms of roles and responsibilities is:**

   _____

   _____

   _____

   _____

   _____

   _____

3. **True or False:** A person's role in process improvement is the same as his or her role in other work.

4. **Which of the following is a good example of a role (may be more than one answer):**
   a. Senior manager
   b. Project stakeholder
   c. Vice President

5. **One of the factors that determines the number of roles you will find in an organization is the _____ of the organization.**

Figure 2.1 Chapter 2: What Do You Think? What Do You Believe?

GP 2.4 comes as close as a model should to being explicit about roles and responsibilities. Its subpractices address assigning responsibility and authority for performing tasks and processes. However, it doesn't tell you how to go about doing this, nor should it. GP 2.7, which addresses the identification and involvement of relevant stakeholders, is my all-time favorite GP. This practice is difficult to implement, yet such a powerful improvement when implemented well. But there's a catch, implementing this GP will be all but impossible for an organization to get its brains and arms around unless roles and responsibilities are defined and being practiced. People simply don't grasp the idea of being a stakeholder unless they perceive that their roles and responsibilities somehow align with the work for which someone has identified them as stakeholders.

Say you're an analyst and your main job, as you know it, is to perform system tests. A project manager comes to you and tells you that he'd like you to participate in peer reviews because you're a "stakeholder." But if you're really busy and under a deadline to finish some testing and someone schedules you to participate in a peer review, which activity is going to get flushed? That's right, the peer review. Why? Because as far as you're concerned, you get evaluated and compensated for doing testing, not for doing peer reviews. However, if your defined roles and responsibilities required you to participate in peer reviews and if you knew that at least a part of your compensation, performance evaluation, or promotion was based on peer review participation, you might make a different choice in the previously described situation. Even if you still make the same choice and blow off the peer review, at least you know the cost of your decision.

The authors of CMMI did the right thing; they did not prescribe to organizations how they should be structured in terms of departments, functions, roles, positions, or responsibilities. CMMI (or any other intelligent reference) cannot do this for you because every organization had different strategies, goals, and business needs and it is those things that should influence the organization's structure. This chapter also does not prescribe a particular definition of roles and responsibilities; it simply provides information on the importance of role definition and how an organization can accomplish this important task.

THE POWER OF KNOWING WHO DOES WHAT

If you question what you see and hear — which, by the way, is generally a good habit to get into if you're in process improvement — you might be wondering why you suddenly find yourself in a chapter about organizational roles and responsibilities and why this topic appears so early in the book. The answer is simple: in almost any systems engineering organization, almost every individual will have some basic understanding of the nature of his or her role, even if they can't articulate their roles very well. People will also believe they know the roles of others with whom they work and they will claim to know the relationships between roles. In both these areas, their knowledge will most certainly be wrong and based on utterly false and fact-free assumptions. Why? Because the nature of the human ego, particularly in Western society, dictates that "I am the universe and the universe is me." What I do matters and what other people do does not matter until it messes with my universe. To quote a friend of mine, "When I close my eyes, it is night."

One of the first things my consulting practice — Natural SPI — does when we engage a client is to try to figure out the answers to the following questions:

- Who does what?
- Why do they do what they do?
- When do they do what they do (in consequential relationship with what other people do)?
- For whom do they do what they do?

Notice we don't ask the question, "how do people do what they do?" which is the prevailing question asked in an appraisal (e.g., SCAMPI) used to baseline the organization's current state. That's because "how" people do their work is trivial at this point in time (the beginning of process improvement) and can't possibly be answered in a permanent way until the first four questions are resolved.

The above four questions are amazingly simple, yet the answers are amazingly complex. We will initially obtain confident answers to such questions. However, in the months that follow, we gradually "peel the onion" and find out that the initial answers weren't even close. It doesn't seem like a problem, at least not a big one. The organization muddles along. Confusion and conflict frequently arise over who is supposed to do what, and these perturbations range in intensity from snippy e-mail notes to all-out yelling, fist-pounding turf wars. Yet somehow in the end, everybody manages to get the product out the door so yippee, the organization gets to survive one more quarter.

Living from one system release to the next by pretending we all understand each others roles and responsibilities might get the organization through software or system releases, but CMMI-based process improvement won't have a prayer of succeeding. In model-based process improvement, consensus definitions for roles and responsibilities is critical to success and that is why it deserves its own chapter. It is also something the organization will want to start working out way before it begins planning its process improvement work (per Chapter 3) since planning assumes a certain level of understanding of roles (like who is involved in planning). First, let's review what has happened in organizations countless times as a way of informing you of what will occur in the future if roles and responsibilities don't get defined.

When Good Roles Go Bad

All of the following scenarios are based on real situations I have observed in my work.

- The system is delivered and a customer complains that it doesn't satisfy their requirements. The customer blames the delivery organization for not understanding the requirements and the delivery organization blames the customer for providing poor requirements. Both sides are right and both sides are tragically and stubbornly wrong. No lessons are learned and both parties simply move their attention to the next crisis.

- The SEPG members perpetually whine about management not being good sponsors of process improvement. The managers don't see "process improvement sponsor" anywhere in their job descriptions or bonus plans and wouldn't intuitively know what being a sponsor of process improvement means under any circumstances. So, the SEPG people must simply be chronic whiners and the senior managers are demonized. Nothing changes, especially the organization's growth and maturity level.

- The organization's testing department hears that SEPG is going to establish a "QA department." The department goes ballistic upon finding this out because, "QA is our job ... why do we need another QA department?"

- Two organizations, each dependent on the other for developing subsystems of an overall system, are constantly at odds with each other over collaborative work. The poison of suspicion and paranoia pervades their communications because each organization believes the other is trying to take away work. There's no good way to objectively resolve the situation because the leadership in each organization "knows" what their roles are; no one can convince them they don't know.

- The project manager and the program manager learn that they are both "responsible" for the project plan. The project plan never gets written because for each person it was "someone else's job."

- A system engineer sends an e-mail note to a vendor that identifies a problem in the system. Three months later, the problem isn't fixed and the system engineer blows up at the vendor. The vendor doesn't understand why the SE is so angry.

- The organization's process improvement group is chartered with implementing ISO 9001. When a SEPG is formed to implement CMMI-based process improvement, the other process improvement group is incensed because the SEPG people are intruding on "their territory."

When we get caught up in situations like these, it's easy to dismiss each of them as a one-off event. Yet when we view multiple similar situations, a pattern forms which allows us to take a system view of the overall problem. One of the primary root causes of all of these scenarios and in many other similar situations is the lack of defined and concurred roles and responsibilities. People in the organization simply assume that they know their roles and the roles of others. Those assumptions are valid except for all the times they are proven to be invalid.

Why Defining Roles and Responsibilities Is So Hard

Defining the roles and responsibilities in your organization is as difficult to do as it is critical to successful process improvement (and all other aspects of business). There are many reasons why this task is difficult, but the primary reasons are:

- There are no universal definitions for roles and responsibilities; there is only infinite variation on a theme.
- We don't have much practice at articulating what we believe to be our roles and responsibilities in the workplace.
- In some organizations, there are many more roles than there are people.
- In some organizations, there are many more people than there are roles.
- Defined roles and responsibilities enable accountability.

Let's take a brief glimpse into each of these barriers to defining roles and responsibilities in the interest of learning how the organization can make progress in this area.

You Say "System Engineer," I Hear "Project Manager"

Until the point at which we enter the professional workplace, who we are and what we do are so integrated at the subconscious level, that there's never any reason to question it or talk about it. What we do *is* who we are. I'm a baby so I cry and drink milk. I later play rough with other boys and tease girls, ergo I am a young boy. I'm a teenager, so I go to school, rebel against my parents, and try to get dates with girls. When I find myself listening to a lecture in a large hall, I must be a college student and that man standing up giving the lecture must be the professor because he is teaching. Even looking around us, we automatically classify who people are and what they do into stereotypes. We see a person driving in a police vehicle in a police uniform — "police officer" is not only what she does, it's who she is. We see a car salesman on TV and the person and his role are one and the same.

So then we enter the professional workplace. I'm a software engineer for one company for several years, so I begin to believe I know what a software engineer does. Then I switch jobs and go to work for another company as a software engineer. My coworkers have an understanding of the responsibilities of a software engineer and I have an understanding of the responsibilities of a software engineer, but the two understandings are different and we don't find out until a situation causes the misaligned understandings to surface. The problem is that the job of a firefighter may be the same no matter which city he protects from fire. But roles and responsibilities of a software engineer, system engineer, project manager, vice president, analyst, or quality engineer are likely to differ from one organization to the next depending on the nature of the business, the market, and the evolved culture of the organization. Then I take a better offer with yet another company as a program manager. I've seen program managers at work, so I think I know what a program manager is and what I'm supposed to do. Yet again, my ideas of a program manager are radically different from those of my employer and once again the harsh light of conflict and reality shines on the misaligned understanding.

What all of this adds up to is the simple fact that there are no universally accepted definitions for roles and their associated responsibilities in software and system engineering organizations. This fact, compounded with the natural human tendency to believe that a job title has the same meaning everywhere, makes defining roles and responsibilities difficult.

I Know What I Do, But Don't Ask Me to Explain It

Another dynamic that inhibits from defining roles and responsibilities is the simple, observable fact that, over time, we become so adept at doing something that performing our job becomes "second nature" and what we do and how we do it moves from our consciousness to the subconscious. Subscribing to one theory on human learning and skill adaptation, all of us more or less take this path:

1. We start out unconsciously incompetent: We don't even know what we don't know. We are neophytes in an area of knowledge and skill and we don't have a clue as to what or how much we have to learn.
2. Next, we are consciously incompetent: We are aware of what we don't know and what we need to learn. We are novices, not yet particularly skilled or learned, but at least we know what knowledge and skill to obtain and how to do so.

3. Then we are consciously competent: We are aware of what we know, how we do our job, etc. We have become proficient and we can articulate our proficiency because the knowledge and skill live in our conscious mind.
4. Finally, we are unconsciously competent. We are experts who are extremely good at what we do. But we are no more aware of our knowledge and skill than we are of our breathing or heartbeat.

When we look at these four evolutionary stages of competency, it seems obvious that if we were to coach or mentor someone in a new area of skill and knowledge, we would use someone in the third stage — consciously competent — since such a mentor would be able to articulate and thus transfer knowledge. But is this what happens? No. What usually happens is that we expect the "experts," those who are unconsciously competent, to mentor novices. And it doesn't work for the same reason I can't explain to people how I write — I don't know, I just do it; it's what I am. Yet 20 years ago, I probably could have explained the process of writing to someone.

More Roles than People

In an idyllic world, there's a job for every person and a person for every job. But in the world in which you and I live, the number of different jobs or roles within a particular organization does not coincide with the number of people in that organization capable of fulfilling those jobs. Additionally, when the ever-swinging pendulum management theory swings from specialization to generalization, we often see organizations in which one person is performing in multiple roles. Finally, the number or roles that evolve in an organization is often correlated to the overall size of the organization in terms of number of employees and the number of different lines of business or markets in which the organization partic-ipates. As shown in Figure 2.2, the definition of roles and responsibilities becomes particularly difficult as you move down in organizational size, because in the smaller organizations, individuals often have many roles and many responsibilities. Figure 2.2 illustrates just a small subset of the plethora of roles you could find in a global company of 10,000 or more employees and how the number of roles gets reduced when translated to a 1000-person organization. Finally, by the time you get down to the ten-person consulting firm, such as my firm Natural SPI, there are two or three people doing everything from running the Board of Directors meet-ing to running a check to the bank for deposit.

Going the other direction — from very small organization to very large organization — you often find such a diffusion (confusion?) of roles that

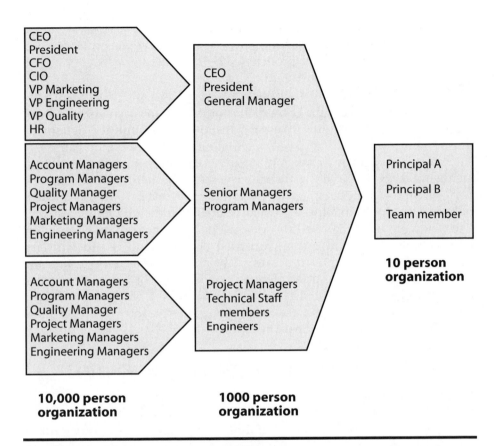

Figure 2.2 Organization Size and Number of Roles

the names of some roles have almost no obvious or intuitive meaning. I know of one very large, global consulting company in which it seems every other person is called a Vice President or Senior Partner. These terms might mean something to the people in this organization, but an outsider, such as a member of an appraisal team, would have a difficult time figuring it out.

More People than Roles

The converse of having more roles than people is, of course, having more people than roles. This one is scary because if the organization does define their roles and then divide them up among the workforce, and there are people left over with no roles, management will draw only one logical conclusion — that's right, some of you have to leave. Less frequently and usually in organizations with a lot of capital to throw around, management will simply invent some roles so that everyone can justify showing up and collecting a paycheck. You can often find people who don't have a real role in something called "special projects."

Fear of Accountability

This one is really the rub. Look, if the organization's management and I decide and concur on my roles and responsibilities and we document those things, then people have something against which they can measure my performance in the job. That's great for performers; but it is anathema to people who really don't want to be held accountable for performing or producing. The postmodern workplace does contain people who would much prefer to have their role defined rather loosely, like "show up every day and do some work." This type of role definition is usually assumed and not documented because it would look pretty silly to write down "show up every day and do some work." If I am among those who really don't want to be measured or held accountable for performing or producing, this is exactly how I will want to leave things. The process focus people are not going to get my cooperation in defining roles and responsibilities.

WHAT IS A ROLE? WHAT ARE RESPONSIBILITIES?

Because it is our nature to assume that the meaning and understanding of words is universal — your understanding of a word or phrase is the same as mine — people in organizations often use the words "role," "responsibilities," "job," "position," and "title" interchangeably because there are no consensus definitions for these words. That's where the trouble begins. Let's say I am responsible for leading CMMI-based process improvement in your organization and I ask you what your role is. You respond, "senior manager." That answer doesn't really do me any good, because I still don't know the relationship between your work and my work.

There are many academic definitions for the terms "role" and "responsibilities," but the definitions your organization should use are those that have relevance to your culture and to which people can agree. For organizations that haven't yet defined these two terms, which probably includes most of the organizations using this book, here are some starter definitions for these two terms which experience tells us work.

Role

A role is a brief, summary description of a person's function in relationship to a particular aspect of the business. Thus, roles are relative to both the hierarchical structure of the organization and a particular endeavor or aspect of business. For example, a systems engineer may have the role of "producer" in relationship to engineering and could have the role of "sponsor" in the organization's CMMI process improvement effort. Hence, the words associated with roles are phrases such as:

- Process improvement sponsor
- Project stakeholder
- Project X manager
- Process user
- System quality assurance
- Requirements analyst
- Software developer
- Purchasing approval authority
- Facilities manager
- Customer service representative

You'll notice that there are a few items in the list above that you have heard referred to as "titles," not roles. That is because all roles can be given titles, but not all titles necessarily imply the role. This is why you will sometimes hear a person who has one title say that she "wears many hats." "Hats" is a reference to the many roles in which the person performs.

Responsibilities

As a starter definition for "responsibilities," the organization can say that they are a summary list of the primary tasks, activities, or type of work performed in a role. So, unlike a role, which is relative to both organization structure and endeavor, responsibilities are primarily relative only to the role, even though the same responsibility may be repeated in different roles. Responsibilities describe the work a person is expected to perform and can be used to measure job performance. Responsibilities include statements such as:

- Review and approve project plans
- Develop process descriptions
- Perform quality assurance audits and report the results
- Plan organizational training
- Deliver project management training to project managers and leads
- Create the project management plan
- Produce project review slides
- Collect, analyze, and report customer satisfaction measures
- Participate in peer reviews as requested by project management

- Attend product design reviews
- Monitor subcontractor performance and report status

Notice the language used in the above sample responsibilities. They all start with an imperative verb, with the implied agent being you in your role. They are also all verbs which explicitly or implicitly generate physical actions or outputs that can be observed or measured. The use of this language is intentional; it is the style of language that gets people to do things. If I documented in your responsibilities the statement "peer reviews are performed," that lets you and everyone else off the hook to actually perform peer reviews. You can just assume that if they're performed, somebody else must be doing them; you were not explicitly directed to perform peer reviews.

The Difference between Titles and Roles

Once upon a time in America you could look at an organizational chart or a list of titles in a company and make some accurate guesses about the roles of people in certain boxes or people with certain titles. John in the sales box was a marketing person. Donna, with a "VP" before or after her name, was told by the president what to do and consequently told the people reporting to her what to do — you know, VP stuff. Organizations were organized vertically in silos, with each business unit or department — accounting, IT, marketing, engineering, manufacturing, quality assurance, etc. — operating as its own fiefdom, almost independent of the other organizational units within the boundaries of the company or larger organization. We all knew what we were (our roles) and what we were supposed to do or not do.

But vertically structured organizations, if they have somehow managed to survive, are today mostly useless sentimentality. The postmodern organization is an integrated system of systems: people (social) systems integrated with systems of nearly intelligent tools (technology) integrated with process systems. These systems and their subsystems — people, processes, and tools — are inextricably interwoven throughout the traditional organizational functions and units. Additionally, people today are required to know more jobs and do more jobs as the need for specialization in people is replaced by specialization in technology or migrates to advanced fields of science. We can no longer make accurate guesses about the role of the person with the project manager title; we can only make reasonable inferences about her role at this moment by observing the work she is performing at this moment.

Traditionally, we have also consciously and subconsciously ascribed characterizations such as power and authority to people having certain titles or appearing near the top of the organizational chart. Though there are still some sectors of American enterprise — namely, the defense industry — in which title comes with power, authority, and influence, this paradigm is also disappearing from the postmodern business landscape. In some industries such as manufacturing, capital still dominates talent in the never-ending struggle between the two forces. However, software and systems engineering is driven by invention and innovation, so talent holds all the trump cards and, when talent walks, the capital follows. Thus, in a software or systems engineering shop, you can find power and influence held by the talent, irrespective of their titles or positions in the organizational chart. In leading edge technologies, talent *is* the capital.

This picture is mostly accurate in Western society and is mostly not accurate elsewhere. I remember sitting in a class at SEI in which about half the students were from Southeast Asia and Japan. In class exercises in which we had to make determinations about decision authority in hypothetical "case studies," the students from non-Western cultures would inevitably point to a name sitting in a top box on the organizational chart as the "obvious" authority and influence figure in the fictitious company. During breaks, I asked these people about their thought processes. At first, they were confused as to how I could be so dense as to not understand that a person high up in the organizational chart or having an "executive" sounding title is always in charge. Once I explained to them that such assumptions were not always true in American businesses, they then knew we were all dense.

DEFINING ROLES

So you know why defining roles and responsibilities is critically important to process improvement and why it is such a difficult task. Don't give up; it's difficult, but not impossible. At some point, people involved in defining roles and responsibilities will start talking to people to find out what they think are their roles and responsibilities. But before anyone starts that activity, people doing this work should first build a design or an architecture for this type of information.

We've already discussed in the previous sections how a person often has one title or position, multiple roles, and then multiple responsibilities for each role. So following that logic, we can construct an architecture that gives us a way to codify the roles and responsibilities. Table 2.1 is one example of such an architecture.

Table 2.1   A Design for Defining Roles and Responsibilitie

Position: System Engineer (Byron Davies)
Role: System Engineer
Responsibilities:
Design/architect end-to-end system including logical, user, and process interfaces between subsystems.
Perform peer reviews of system architecture with relevant stakeholders, incorporate feedback, and obtain concurrence on delivered architecture.
Attend system development reviews and provide system engineering perspectives.
Manage system development and delivery risks as assigned by program manager.
Role: SEPG Member
Responsibilities:
Participate in all SEPG meetings.
Develop, revise, and review process definitions and other process assets as assigned by SEPG chair.
Provide system engineering knowledge and perspectives to all SEPG work and outputs.
Role: Employee Management and SE Administration
Responsibilities:
Work with direct reports (engineers and architects) to establish and document semiannual individual performance objectives.
Provide coaching, mentoring, and performance feedback (against approved objectives) to direct reports as needed.
Provide formal (documented) semiannual performance feedback and rating to direct reports.
Based on performance and in accordance with HR policies, administer direct report promotions and compensation.
Work with direct reports to establish and document individual training plans.
Role X: Additional role description
Responsibilities:
Responsibility *n* for Role X
Responsibility *n*+1 for Role X
Position: B (Sally Somebody)
Role A for Position B: Additional role description
Responsibilities:
Responsibility *n* for Role A
Responsibility *n*+1 for Role A

Where to Start

Once a design or architecture is established for defining roles and responsibilities, it's time to start populating that design with live information. The process focus people or process team responsible for defining roles and responsibilities should pilot their design by defining some process focus roles first before trying it out on other people in the organization. This will give the designers/definers a chance to find out what works and what does not with the work products they intend to use to define roles and responsibilities and mitigates the risk of them unnecessarily irritating other people in the organization with a bad and unproven design.

The next step is to take great precautions to avoid yet another common mistake made by well-intentioned process people: defining roles and responsibilities based on academic or textbook definitions or worse, based on myopic views of the business world resulting from limited experience.

For example, an intelligent but inexperienced person responsible for defining roles and responsibilities would quickly determine that he can simply borrow this definition for "project manager" from CMMI:[2]

> ...the person responsible for planning, directing, controlling, structuring, and motivating the project. The project manager is responsible for satisfying the customer.

However, this definition is intentionally broad such that it generically describes every project manager in every systems organization. If someone puts this out into your organization as the definition for project manager, it will have one of two consequences:

1. People who are not project managers will think the definition describes their role. Based on their current position and role, they will either think they just got a promotion or a demotion.
2. Some people who thought they were project managers will not see themselves in the definition and will be either elated or upset that they are no longer a project manager.

The definition for project manager (and all other roles) in an organization must be the definition that describes that role in the context of that organization and its culture. In some cases, an organization has evolved its own labels for roles and there is no reason to change the terminology. For example, the different terms for "project manager" I've seen are project lead, software lead, lead engineer, task manager, task lead, and team lead. In each of these organizations, the definition for the role of people doing project management were similar, but they chose different labels for that role. Sometimes in these situations, a SEPG member will insist that the people in the organization change their language, as if there is some secret law that says organizations must structure themselves in accordance with textbooks and other literature. As change agents, people who pursue this route have almost no chance of success. Who cares! Let your organization assign to its roles the labels to which people are accustomed. It is the definition of those roles and the corresponding responsibilities that matter.

To begin defining the roles and responsibilities, it is critical to first realize that many of the organization's roles exist in practice and that the task at hand may be one that is more documenting the existing roles rather than defining them. However, if the organization is just starting out with CMMI and process improvement, assuming that roles are accurately defined and practiced is too risky; someone needs to go find out. It will take a little bit of detective work, but it will be worth the effort to avoid inventing roles that stand a good chance of upsetting people and arming them with justification for why the process improvement is a waste of their time. This is where positions and titles come in handy. There are job and position titles which are almost universally ascribed to groupings or types of work. Table 2.2 lists some typical job and position titles you can find in most software and systems organizations, the type of work people in these positions typically perform, and the CMMI process areas correlating to that type of work.

CMMI Process Improvement Roles

Perhaps before someone starts defining the roles and responsibilities for everyone in the organization, it might be prudent to first begin defining the roles and responsibilities needed for the process improvement project. From SEI's Managing Technological Change (MTC) course,[46] we know there are three primary roles in any organizational change, including CMMI or CMM-based process improvement:

1. Sponsor
2. Change agent
3. Target (participant or "victim")

MTC elaborates on the responsibilities for these process improvement roles, but it still doesn't paint the whole picture because roles are fluid and situational. I may be in the role of a sponsor in one endeavor, but a change agent in another situation, and a target in yet another situation. Meanwhile, I am also a SEPG member and a requirements analyst. And that's all just my work-life roles. Once I get home, I'm might also be a spouse, a mother, a steward and protectorate of domestic animals, and a

Table 2.2 Where to Look for Roles in the Organization

Typical Positions	Type of Work Typically Performed in these Positions	Related CMMI Process Areas or Practices
Engineer Software Engineer Developer Analyst Programmer Designer Software Developer Architect System Engineer Technical Staff Member Technical Specialist	People having position titles or job descriptions similar to these are usually the people involved with analyzing and defining requirements, designing the system or products, developing software or software-intensive systems, and maintaining (fixing defects) the systems and products they support. People in these positions also are commonly involved in integrating and testing system components.	Requirements Development (RD) Technical Solution (TS) Product Integration (PI) Validation (VAL)
Project Manager Project Lead/Project Leader Software Lead System Engineer Senior Engineer Program Manager Team Lead Task Lead/Task Leader Manager Programming Lead IPT Lead	People with these jobs or position titles are usually individuals who have moved into positions with some level of authority to provide leadership or guidance to other workers. These positions are frequently involved in estimating and planning work and then managing the work performed by others and reporting the results.	Project Planning (PP) Project Monitoring and Control (PMC) Risk Management (RSKM) Integrated Project Management (IPM/ IPPD) Generic Practice 2.2
Quality Assurance Quality Control Quality Manager Quality Specialist Quality Engineer	Although traditional jobs and positions related to quality have been primarily associated with product quality, these people are natural candidates to take on the role and responsibilities for assuring process quality also.	Process and Product Quality Assurance (PPQA) Validation (VAL) Generic Practice 2.8

(continued)

Table 2.2   Where to Look for Roles in the Organization (Continued)

Typical Positions	Type of Work Typically Performed in these Positions	Related CMMI Process Areas or Practices
Configuration Engineer Configuration Manager Configuration Specialist Data Manager Technical Publications Manager Integration Engineer Change Control Board member Configuration Control Board member	People with titles such as these are involved in managing and controlling hardware or software configurations or managing documentation versioning. There are people who argue that there is a difference between change management and configuration management, but they are wrong. Configuration management is merely a subset of the larger change management.	Configuration Management (CM) Generic Practice 2.6
Purchasing Agent Acquisition Manager Procurement Specialist Buyer Contracts Manager Contract Officer Accounts Payable personnel	People in these positions regularly deal with vendors and suppliers to procure materials and services for the organization and for projects. In the government and defense sectors, personnel in these positions usually follow rigorous procedures due to their requirement to be compliant with Federal Acquisition Regulations (FARs) and public law.	Supplier Agreement Management (SAM) Integrated Supplier Management (ISM)

little league football coach. No one gets to be just one role all day, not anymore.

In CMMI process improvement, it is no coincidence that people's roles will closely resemble those you will find on a system engineering project team (read "Establishing the Process Improvement Project Team" in Chapter 3 — Managing the Process Improvement Project). The organization will need to define process improvement roles which encapsulate what people do in the process improvement project. Hence the role "SEPG member" only tells us that a person attends SEPG meetings and we know that processes don't get defined and implemented simply as a result of people attending SEPG meetings.

The critical roles and responsibilities which need to be defined for a CMMI-based process improvement project are:

- *Customer:* If an organization cannot define a customer or customers for process improvement, that is, they cannot define who wants the processes improved and why, then it should reconsider engaging in an activity as expensive as CMMI-based process improvement. The customer or customers of process improvement can be people within the organization using CMMI and they can be people outside of the organization. The customers of CMMI or improvement project are also probably relevant stakeholders in many of the process improvement activities.
- *Project manager:* This is the person or persons who have the ultimate responsibility for planning the process improvement project, obtaining resources to do the work, assigning tasks, and then managing the project in accordance with plans. This person or people can also be the leader of the process focus team (i.e., SEPG), but is not necessarily so.
- *System engineer:* The person or persons responsible for designing the overall process system and ensuring the successful integration of its subsystems and components. This person typically leads the establishment of standards and processes for developing the organization's process assets and also designs the organizations process repositories such as a Process Asset Library (PAL).
- *Engineers:* People responsible for designing and developing the organization's processes and process assets.
- *Educators/trainers:* People who possess the skill and knowledge to transfer knowledge to others.
- *Analysts:* People who know how to measure and analyze the efficacy of processes in the organization.
- *Configuration management:* By default, the organization's process focus group is usually the "change control board" for the organization's processes and process assets. However, this should not be the default assumption since configuration and data management is a learned skill and not one always possessed by SEPG members.
- *Quality assurance:* The process improvement team will (or should) be operating and conducting its activities in compliance with defined plans, processes, and standards. The project needs someone to provide objective verification that the project is executed in compliance with those governing plans, processes, and standards.

DEFINING INTERORGANIZATIONAL ROLES

If the organization's leadership and process focus people are strong, tenacious, and courageous, the organization can eventually define, implement, and institutionalize roles and responsibilities for individuals. Doing so will reap tremendous, measurable benefits in operational efficiencies and also progress the organization forward in terms of CMMI process capability and maturity. The organization can stop there, but the job is really only half complete. The real benefit comes from defining interorganizational roles and responsibilities and this task is much more difficult than doing this work for individuals within one organization.

In the postmodern world of business, there really is no such thing as an organization operating independently of other organizations. Some people maintain the illusion of the organizational "island," but it doesn't exist. In our world today, no matter what type of business we're in, there are almost countless relationships between our organization and others. Most of these relationships can be characterized using words such as dependency, constraint, provider, customer, codeveloper, and collaborator. And it is possible to identify, define, and get agreement on the interorganizational roles and responsibilities; it's just very hard work and that's probably why most people don't do it.

The rewards of defining interorganizational roles and responsibilities are pretty good, but the cost of not defining them is high. In my consulting business, we experienced a classic case of the high cost of poorly defined interorganizational roles. We were consulting two different organizations, both which were providing some similar engineering services to a common customer. In isolation, each organization drew the system engineering life cycle and then divided up the life cycle phases according to what each organization thought was their domain and responsibility. Guess what happened when someone finally got copies of both mapped out, cut-up life cycles and looked at them side by side? Yep — anger, defensiveness, protectionism, fear, and offensive posturing. "How dare they say they do X ... X is our job!"

Why do organizations practice this type of territorialism? You know the answer: money. In the case cited above, the stakes were high. The more work each organization could lay claim to, the more they could bill the customer. Who doesn't try to expand their business in this manner? Fortunately, there are many other ways to expand the business — innovation, increased value, increased service — and all of them are easier and more cost-effective than fighting with another organization over a piece of the pie. The lesson learned is that when organizations define their role and responsibilities independently of other organizations with which they have relationships, one of two situations will eventually occur:

1. Two or more organizations will lay claim to the same work, causing time and money wasted on political fighting.
2. None of the codependent organizations will claim responsibility for a piece of work and that work will not be performed.

The relevance and importance of interorganizational roles and responsibilities to CMMI-based process improvement are that the definition of such roles determines in large part how an organization defines its processes and standards. Let's say that an organization defines its processes for requirements management with the assumption that the customer is responsible for clearly defining the requirements before giving them to the system engineering organization. The requirements perpetually come into the system engineering organization poorly defined. The engineering organization does its best to define the requirements, but usually ends up not satisfying some of them with the deliverables. The customer complains and the engineering organization tells the customer it's the customer's fault for not clearly defining the requirements. The customer claims that requirements definition is not its responsibility. The leadership in both organizations feels that it's more important to be right than to fix the broken relationship, so the situation continues release after release.

Interorganizational roles and responsibilities can even be viewed from within a single organization. The processes say that the Project Management Office (PMO) is responsible for project planning. When representatives from PMO ask engineers for help in estimating work, the engineers say, "planning is your job; I just follow your plans." Inevitably, PMO tries to put together plans without estimates from experts and then the technical team complains that the plans are unrealistic.

There are some guidelines an organization can put in place that can mitigate the high cost of ill-defined interorganizational or interfunctional roles and responsibilities.

Collaborative Roles and Responsibilities Definition

When your organization begins defining its internal roles, responsibilities, and processes, invite representatives from other organizations with which your organization has relationships to participate. Tell them they are stakeholders in your organization's internal processes and standards and be ready to explain to them why and how they are stakeholders. Let them know that your organization doesn't want to unilaterally define roles, responsibilities, and then processes that might somehow affect their work without getting their input. Sometimes you'll get takers using this approach, but sometimes you won't. At any rate, you will have the clear conscience (and record) that you asked.

Identify and Work toward Common Goals

One proven way to bring two or more organizations to work together in collaboration is to find or define common goals or objectives which would not be obtainable by any one organization. If another organization understands that its success in a particular endeavor is dependent on the success of your organization, then the two organizations are far less likely to work against each other because doing so would be counterproductive to the success of each. For example, let's say a systems engineering organization and a testing organization ostensibly work together to deliver a system to a customer, but there is always strife and finger-pointing between the two organizations. The testing organization misses defects, but blames the engineering organization for not implementing the requirements. Engineering points fingers at the testing organization and says the testing is inadequate. To resolve this, the two organizations could form a joint working group to codevelop the test cases. This way, when things work and defects are removed in testing, both organizations get the credit. When things don't work, each organization has mutual responsibility for the failures. The successes and failures of the two organizations are now inextricably codependent and their mutual fate is based on continually maintaining the health of the symbiotic relationship.

Get Your Own House in Order First

Let's say your organization has done "due diligence" in trying to get other organizations to work with it to collaboratively define roles and responsibilities or has attempted to establish mutual goals with other organizations, but the other organizations just won't play. Your organization still has a goal to improve its processes and that's exactly what it should do. In this situation, the organization's process focus and process definition people should just go ahead and define the standards and processes as best they can with the available information. At a point where a process connects to an external organization, the process definition should clearly and overtly state that it continues into unknown territory in another organization. Document the requests for participation and the refusals to participate from other interfacing organizations. At some point in the future, someone is going to ask all involved why they're not working together. Your organization needs to at least be able to show that it tried and that it had no choice but to make progress without input from the connected organizations.

DO'S AND DON'TS

Now that you've read about the importance of defining roles and responsibilities and have hopefully learned some things you can implement in your organization, here are some but not all of the lessons in abbreviated form.

Do

- Understand the difference between roles and job titles or positions. A person having one title may have multiple roles depending on the activity in which she is engaged or depending on a particular situation.
- Before defining roles and responsibilities, establish an architecture or design for documenting them. This will ensure that you codify the critical information for each role defined.
- Learn about the job titles or position titles currently in use in your workplace. They can serve as clues to finding people in the organization who match the roles as they are defined in standard industry terminology.
- View implementing CMMI-based process improvement as an engineering project to develop a "process system." When viewed this way, the roles specific to process improvement are similar to the roles needed to successfully plan and execute a system engineering project.
- Identify and define people's roles and responsibilities because it is a critical factor in successful CMMI-based process improvement.

Don't

- Don't assume that peoples' job and position titles will describe what they do. Titles are legacy conveniences that have lost meaning through overuse.
- Don't assume that roles (or even job titles or positions) have universal meaning. They vary depending on the size of the organization and the nature of the organization's business.
- Do not think that the organization will go very far with CMMI-based process improvement without defining roles and responsibilities. Failure to accomplish this as one of the first activities of process improvement will result in extensive waste and rework because people will not know who is supposed to do what.

WHAT DID YOU LEARN? WHAT WILL YOU DO?

Now take the post-chapter quiz (Figure 2.3) and think about what you've learned and how some of your views toward CMMI-based process improvement have changed. Think about what you will do with the information you've learned (and how it makes you feel).

1. **True or False:** You can usually figure out people's roles and responsibilities just by looking at the organization chart.

2. **Which of the following is not a good example of a role (may be more than one answer):**
   a. Senior manager
   b. Process improvement sponsor
   c. Purchasing approval authority
   d. Vice President

3. **Which of the following is a good example of a role:**
   a. Senior manager
   b. Project stakeholder
   c. Vice President
   d. Engineer
   e. CEO

4. **A person's role is relative to two things:**
   1) the _____ of the organization in which the person works, and 2) the type or aspect of _____ in which the person is performing.

5. **The most important idea I learned from this chapter is:**

   _____

   _____

   _____

6. **I will apply this idea in my process improvement work by:**

   _____

   _____

   _____

Figure 2.3   Chapter 2: What Did You Learn? What Will You Do?

3

MANAGING THE PROCESS
IMPROVEMENT PROJECT

It is not the same to talk of bulls as to be in the bullring.

— **Spanish Proverb**

WHAT DO YOU THINK? WHAT DO YOU BELIEVE?

Take a minute and answer the questions in Figure 3.1. Then, once you've finished reading this chapter, take the quiz in Figure 3.15 — "What Did You Learn? What Will You Do" — to find out how much this information has helped you with your own CMMI-based process improvement. Remember, these quizzes are like process improvement work, there is rarely a right or wrong answer; there are only answers that best suit your organization's business needs.

This chapter is written primarily for people in the organization who have direct responsibility for planning, leading, tracking, and reporting CMMI-based process improvements. This audience includes members of SEPGs or EPGs, process improvement managers, and the like. There are some tips for executives and senior managers, project managers, and engineers and other technical people.

Here's one of the biggest ironies about the process improvement efforts I've been involved with. The people responsible for managing the process efforts love to point out when management doesn't "walk the talk": managers talk about process improvement, but their actions say otherwise. Yet, how many times have you seen people managing process improvement *not* practice what they preach?

71

1. **True or False:** Process improvement or achieving a maturity level is not a project.

2. **True or False:** When senior management sets a deadline for achieving a maturity level, this is an indicator of management commitment.

3. **When process improvement efforts fail to achieve their goals, it is primarily due to:**
   a. Lack of management commitment
   b. Uncooperative people
   c. Poor project planning and management
   d. No tools available
   e. None of the above

4. **True or False:** The SEPG creates and implements processes, but doesn't have to follow any.

5. **Requirements traceability applies to which of the following:**
   a. Software development projects
   b. System development projects
   c. IT shops
   d. Process improvement
   e. Virtually any undertaking for which the customer can be identified

Figure 3.1   Chapter 3: What Do You Think? What Do You Believe?

Here's what I mean: the CMMI process areas REQM, PP, PMC, IPM and RSKM are all about managing requirements and planning and managing system projects, right? So why should the project called "process improvement" or "CMMI implementation" be any different? Why do those of us in SEPGs or similar process focus units, the people who are supposed to manage and control process improvement, often not do any of the following?

- Gather and manage requirements for implementing CMMI
- Estimate and plan the process improvement efforts

- Develop a process improvement project plan
- Conduct regular tracking and oversight of the process improvement project
- Monitor stakeholder involvement and commitment to the process improvement's success
- Have someone objectively verify our work against our own plans, standards, and procedures
- Make sure our process work products are under change or configuration control
- Peer review our work products
- Plan integrated team work and collaboration
- Measure and collect data on the process improvement effort

Why should we manage model-based process improvement any differently than the way we ask system development and engineering managers to manage their projects? The answer is, we shouldn't. If we do, we are hypocrites and not credible to the rest of the organization. Let's take a closer look at how you could (and should) manage your process improvement effort as a project using the same CMMI-based processes you're going to ask the system projects to use.

If you're an executive or senior manager in the organization and you really want the CMMI process improvement effort to succeed, don't treat it any differently than you do other high-priority programs or projects in the organization. Let the people managing the process improvement effort know that you expect to see fact-based estimates and plans (documented) for achieving the process improvement goals. Ask for the same kind of progress, status, and measures that you ask for from other programs or projects. Base your decisions regarding resource allocations to the CMMI project using the same business-based criteria that you use for systems projects.

You'll have to give some also. Realize that demanding a completion date (e.g., achieving a maturity level) for the process improvement project (or for any type of project) that is not based on historical or industry data or fact-based estimates is a low maturity behavior and won't inspire confidence among your people that you really buy into the principles of process discipline. If you "time-box" the CMMI project because that is how you manage all other projects, accept that you've increased risk, descoped the deliverables, will have to increase allocated resources, or all three.

Finally, don't allow yourself or others to treat the process improvement project as if it is in addition to the "real work." You will hear people grumble things like, "I can't do this process stuff because I've got 'real work' to do." At times, you'll feel the same way. Try very hard to change this view for yourself and others. Improving the way your people work is as much a part of the job as coding, designing, testing, etc.; don't let CMMI or process improvement become extracurricular.

THE MODEL AND THE REALITY

Both SW-CMM and CMMI provide excellent guidelines for planning and managing projects. It's too bad that lots of people involved in process improvement work fail to see the relevance of these guidelines to their work. But even if we just focus on software-intensive systems, CMMI does not or cannot address many of the realities of project planning and management. CMMI works as designed because it was never intended to be used by organizations as a convenient substitute for thinking.

Some of the mushy, sticky, gray areas of real project management for which you won't find an easy solution in CMMI include:

- Inability to establish a consensus understanding of the word "project" (see "Establish a Common Language" in this chapter).
- Determining the origination of a project and figuring out which comes first, requirements or planning. If you don't have requirements, you technically don't have a reason to plan, but don't you need to plan requirements activities?
- Implementing processes and procedures using tools. It's easy to say, "develop a WBS for the project," but almost all of us will be required to implement that task using some sort of tool. Whether the tool is as simple as writing the work breakdown structure (WBS) tasks on a whiteboard or as complex as using an enterprisewide project management system, there will be dependencies and constraints inherent in the use of the tool that will make the actions more difficult than the words.
- Prioritizing work issues. In a perfect world as seen from the view of a model, there are a group of people who work on the project team and they work on that one project until it's finished. While they work on the project, they answer to the project manager. In the real world, people are rarely dedicated to only one project. People must spread their time out over dozens of projects, tasks, and organizational activities. People report to many bosses for the different work they perform and people usually get very little guidance from anyone on how to prioritize their work because each of the bosses tell them that their work has the highest priority. CMMI won't help you much here.

This chapter won't prescribe any answers for these situations because there is no one answer. There are lots of possible solutions, all of which depend on the specific organization's strategy and business model. What this chapter will do is provide experience-based knowledge of how to plan and manage a project for implementing CMMI-based process improvement. This information represents many lessons learned from multiple process improvement projects.

GO WHERE EVERYONE HAS GONE BEFORE

If we make the assumption that project management practices and methods help people manage software and system projects effectively and efficiently — and we must believe this since we're so good at expecting project personnel to employ such practices — then it's logical to also assume that project management practices and methods can be used in any endeavor that meets our definition of a "project." In fact, people who use the principles and practices defined in PMBOK[16] assume this axiom to be true since it is used by managers of many types of projects from building construction to oil exploration to software and systems development.

So the next question is, "what do we mean by the term 'project'?" There are many published definitions for a project and you certainly shouldn't reinvent the term. However, a commercial-off-the-shelf (COTS) definition for project may have to be tweaked to meet your organization's needs. (Don't worry, there are no Definition Police running around arresting people for modifying the meanings of words and phrases used in the public domain!)

The two most applicable places to look for a COTS definition for "project" are the CMMI Glossary and PMBOK. But no matter where you find your definition, you'll find just about everything ever called a project met the following four criteria:

1. It has a start date and an end date.
2. It will deliver something that can be defined, measured, or tested.
3. It will be allocated defined resources.
4. A customer or customers can be identified for the thing to be delivered.

By this definition, every CMMI-based process improvement effort is or should be classified as a project. There is always a start date, or at least one can be defined. There is always an end date, usually a date for achieving a maturity level dreamed up for you by the organization's executives or senior managers. Resources (almost certainly not enough) will be allocated to process improvement because otherwise nothing occurs: processes do not magically materialize. The thing you will develop and deliver is a system: a process system. Developing and delivering a process system is not unlike developing and delivering an e-mail system for internal use — both systems are intended to help people in the organization work more effectively and efficiently. And finally, you and others better know explicitly to whom you are delivering process improvement because — in the words of my business partner and mentor — if you don't have a customer, you should just go home.

So, if a process improvement effort to achieve a maturity level, process capabilities, and measurable improvements looks like a project, feels like a project, and smells like a project, then it is a project. As such, your organization's CMMI-based policies, processes, and procedures for systems engineering apply to the CMMI project as well as they do to the software and systems projects. If you accept this, the good news is that the process improvement project can play on the same field with the other projects in the organization. The bad news is that the process improvement project runs out of excuses for failure.

But wait! There's more good news. When you shift your paradigm and start viewing the CMMI or process improvement effort as a project, you realize process improvement success is not luck and there is a proven path for success: project management. The rest of this chapter tells you and shows you how to apply standard, proven project management techniques and methods to your process improvement project to manage risk and ensure success. "Boldly going where no one has gone before" might be a good motto for science fiction, but it's not an intelligent approach to CMMI-based process improvement. Wisely go where everyone has gone before.

 As a project manager or lead (or software engineer who's smart enough to not accept management positions), you have a lot of experience in successfully managing projects, even ones with impossible deadlines and ridiculous constraints. Too often, people assigned with process improvement or CMMI responsibilities do not have the depth of project management experience you have.

Help your organization's process people understand that their chances of success in reaching the targeted maturity level and other goals will be much better if they apply some project management discipline to the effort. There are dozens of reasons projects fail and a process improvement project is every bit as susceptible to the same sources of risk and failure as software and systems projects. Offer to help the process people estimate, plan, and manage the CMMI implementation as a project. There might initially be some resistance to this concept, but they will understand the need once upper management starts asking for reports on status and progress of the work against plans.

Also, the best way to get the process people to understand if a procedure will work is to first try it out on the process improvement project. For example, if they develop a procedure for gathering and analyzing project requirements, challenge them to apply the procedure to gathering and analyzing requirements for process improvement. It's often a real eye-opener for process people when they try to use their own procedures and work products on their efforts.

ESTABLISH A COMMON LANGUAGE

I had a client who hired me to give them consulting on their enterprisewide CMMI-based process improvement project. After conducting a baseline appraisal (SCAMPI Class B) and establishing their initial process improvement plans, they were anxious to form working groups to start developing the common processes. I tried to get them to do something different: define the words and phrases that make up the core language of process development implementation. They chose not to follow my advice. After all, everybody knows what we mean when we use words such as "project" or "process" or "organization," right?

In a word, no. Conduct this experiment: Go to any five people you work with and ask them to independently write down on a piece of paper what organization they belong to. To prove my theory to myself, I did this experiment when I was at CSC. The five people who played along were all in the same organization at some level, but here are the five answers I received:

1. Julie's department
2. Long Beach
3. Application Services Division (ASD)
4. Quality Assurance
5. CSC

All five respondents were absolutely correct, yet in terms of defining the organizational scope of a process improvement project, the five answers were hundreds of thousands of dollars apart from each other.

You'll find the same to be true for other key words such as "project." Ask any five people what a project is and you'll get two to five different definitions. Language and definition seems trivial to most; surely everyone knows what I mean when I say X, right? That is the way most of us think anyway and we are wrong; language is not trivial and it is critical to success. It matters because if the scope of your CMMI-based process improvement applies to "system" "projects" within your "organization," you and others better have a consensus understanding and agreement on what these and other terms mean.

The words you use make a huge difference in the cost, schedule, risk, and quality of your process improvement project. There are dozens of key words and phrases, but the ones which have the most impact on scoping, planning, and implementing CMMI-based process improvement are:

- Organization (or organizational unit)
- Approval
- System
- Project
- Policy
- Process
- Procedure
- Work product
- Standard
- Senior management
- Stakeholder
- Measure, metric, and measurement
- Requirement

If you're just starting out, refer to Critical Factor 3: Define the process language in Chapter 5 — Five Critical Factors in Successful Process Definition — before you go any further. If your organization's CMMI or process improvement project is already well underway, you should make the time to verify that a consensus understanding of these terms has evolved in the organization. Common definitions for these terms will also be critical to your organization's success in a SCAMPI appraisal.

DETERMINE THE STARTING POINT FOR CMMI PROCESS IMPROVEMENT

The starting point for a CMMI process improvement project consists of two parts:

1. Things the organization already has in place.
2. Things the organization does not have in place and needs or wants.

The organization must have information on both aspects to form an accurate picture of the starting point for process improvement and an accurate starting point is critical to developing realistic plans. Finding out what the organization already has in place is addressed in the following section, "What the Organization Already Has." Figuring out what it needs follows that in the section, "What the Organization Needs or Wants (but Does Not Have)."

Here's some bad news: Understanding an organization's true, real, or exact starting point for process improvement is not possible. The accuracy of the starting point is relative to the best information that can be gathered at a point in time. The accuracy of the information you gather is a function of how much money and time the organization can spend in gathering it and the skills and expertise applied to that discovery. Your characterization of the organization's process capability or organizational maturity changes with time, not only because the processes and their implementation are constantly changing, but also because you are perpetually discovering things about the organization that change the original characterization. Inferences or assumptions made based on things observed last year turn out to be false in the light of newly discovered information this year. (The information was there last year, you just didn't see it.) As more information about an organization surfaces, the context of original observations also changes. What once was fact is no more. Something else you need to know but may not like to hear is that organizations' process capability or organizational maturity can diminish more easily than it can improve and few people will even notice.

From a process improvement planning and implementation perspective, it's important to understand or at least acknowledge the relativity of the "starting point." Remember that your process improvement plans are "realistic" based on an accurate picture of the then-current state. So if the organization's now-better-characterized-current state differs significantly from the then-current state, the plans must be revised accordingly. This means that the process improvement plans are fluid and constantly changing, which again is not all that different from the iterative planning for software and systems projects.

If a moving starting point and a moving target drive you mad, you might consider a different line of work. At least that has been my thought several times as a consultant. In our TARIF client, we conducted a fairly thorough baseline appraisal [at the time, equivalent to the Standard CMMI Appraisal Method for Process Improvement (SCAMPI Class B)][3] to ascertain the organization's process implementation. One of the areas in which the organization initially appeared to be strong was project monitoring and control. Direct evidence was corroborated by indirect evidence and affirmed via interviews. The story of monitoring and control in TARIF looked and sounded coherent and consistent and we based our initial judgments and plans accordingly.

In the ensuing year that followed the baseline appraisal, we would periodically witness behaviors or artifacts which *should* have made us challenge our initial characterization of PMC in TARIF but did not because

— like everyone — we wanted to be right. But when enough facts fly in the face of a theory, the theory loses unless you're insane. We finally had to reset our view of PMC in the organization. Practices that we had deemed implemented (colored green in charts) turned to not implemented (colored red). You'll always have some explaining to do when charts turn from green to red. We had to go to the client and admit that we knew now what we didn't know then and that the plans had to change. We had to admit to being smarter today than we were yesterday and that we might be smarter tomorrow than today. If a moving starting point and a moving target drive you mad, you might consider a different line of work.

Using Appraisals to Determine the Current State

One of the more useful models to come from SEI is the life cycle for process improvement known as "Initiating, Diagnosing, Establishing, Acting, and Leveraging" (or Learning), which is commonly referenced by its acronym, IDEAL[SM].[21] Figure 3.2 shows a picture of the IDEAL Model.

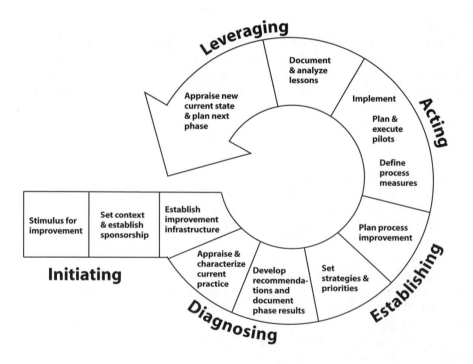

Figure 3.2   IDEAL Model

Many organizations use appraisals as the means to implementing the Diagnosing phase of IDEAL. Appraisals, particularly those that conform to a published methodology such as SCAMPI, are usually effective at yielding an accurate characterization of an organization's current process capability or organizational maturity. However, appraisals, particularly SCAMPI Class As, can be very expensive and highly disruptive to the organization's core operations.

SCAMPI appraisals can be conducted in one of two modes: (1) Discovery or (2) Verification. In Discovery mode, the primary objective of the appraisal is to find (hence, "discover") in the organization, processes and evidence of the use of those processes, and map that evidence to the practices in CMMI. A Verification form of SCAMPI assumes that the organization has already gone through discovery and has already mapped evidence of the use of processes to CMMI. It also assumes the organization possesses indigenous knowledge of CMMI and how to implement CMMI-based processes.

The default mode seems to be Verification and most process improvement consultants I know, such as SEI-authorized lead appraisers, will try to talk you into a Verification form of a SCAMPI appraisal. The cynical side of me wants to believe that it's because Verification appraisals are the quickest and most profitable in this business, but let's pretend there's no basis for such cynicism.

There are many important factors that should be considered when making a business decision as important as conducting appraisals. Like most other decisions in process improvement, the only right answer is that which is right for your organization. Remember that, besides a SCAMPI appraisal, there are other ways of accomplishing the goal of gathering factual process information from your organization. The other ways are described in the following sections. Also, before you hire a consultant or vendor to advise you, read Chapter 6 — Acquiring Process Expertise and Tools.

What the Organization Already Has

Let's just accept for the sake of discussion that, even though you've tried to convince your management to set business goals for the process improvement effort, what your bosses really want you to do is "implement CMMI." Oh, what to do, what to do … so many practices to implement and so little time! You still need to know where to start. Are you simply going to assume that your organization isn't doing anything that relates to CMMI and that you're going to have to plan and implement all of the practices? If you do, you could be wasting a great deal of your time and effort and the organization's money.

It would be extraordinary to find a software or systems organization that has been around for any amount of time that isn't already doing things that could be mapped to CMMI practices. Good software and systems management practices and business practices naturally evolve in most software and systems organizations. So, before you begin planning tasks, before you spin people up to start writing lots of policies and procedures, sit quietly and consider the possibility that the journey has already begun, and that you're simply not yet aware of that fact. Read Chapter 1 — News Flash! There Is a Level 1! That information will help you understand the nature of determining the real starting point for process improvement in your organization.

Call it what you will — gap analysis, evaluation, assessment, appraisal, etc. — you need to perform some kind of discovery and analysis to find out what things your organization is already doing and how those things map to components of CMMI. The analysis of where your organization stands against CMMI can be rigorous or less rigorous, expensive or relatively cheap.

One of the first things you should do is find out if the organization has recently (within the past three to six months) conducted some type of method-based appraisal such as a Software Capability Evaluation (SCE^SM),[48] a CMM-Based Appraisal for Internal Process Improvement (CBA IPI),[5] or a SCAMPI. If you do find that some type of appraisal was recently conducted, and if the organization has not changed dramatically (reorganized) since that appraisal, you can reasonably assume that the results of that appraisal are still valid and can be used for process improvement planning. If more than six months have lapsed since your organization's last appraisal, if it has never conducted one, or if it has recently undergone significant changes, you reduce the risks to your project by assuming that you need to get a fact-based understanding of the organization's current state with regard to process implementation. An appraisal to determine the organization's current strengths and weaknesses in the context of CMMI is typically referred to as a "baseline appraisal" because it establishes the baseline or starting point for process improvement work.

If the organization can afford the cost and interruption of planning and performing a SCAMPI appraisal, then that is what it should do to get the most comprehensive characterization of the current state. Alternatively, you and perhaps a few other people can appraise your organization as a routine, ongoing activity that we call in this chapter, "Everyday Appraisals."

People have often asked me, "when should our organization get appraised?" My answer is always the same, "every day!" I encourage organizations to do what I did as an employee of organizations involved in process improvement: devote a little time each day or each week to collecting evidence of process instantiations and mapping them to the CMMI practices.

In almost any role in any company in the postmodern world, you will receive from 10 to 60 e-mails each day. Inevitably, some percentage of these e-mails or their attachments will be either direct or indirect evidence of the instantiation of a process either at the project level or an organization level. But most people don't hang on to such items. Instead, they start collecting and organizing such evidence about three months prior to a scheduled appraisal and such activity is often hectic and spastic because the organization is in a "crunch" to prepare for the appraisal. It doesn't have to be that way.

I recall one year in CSC during which I was the project manager for the process improvement project of an organizational unit. Starting about two weeks before a CBA IPI for CMM Level 3, senior managers would frequently pop into my cube and ask why I wasn't frantically preparing for the assessment. Weren't there any emergencies? Weren't there any fires to put out? Why were people not scurrying about doing things to get ready? Why weren't people pulling "all-nighters?" Why was there no panic? My calm — the lack of quivering and knee-jerking in my demeanor — made the bosses fret even more.

Well, my manager at the time (who later became my business partner) and I had decided long ago that we weren't adrenaline junkies and that crises were just one of the many manifestations of poor planning, poor organizational skills, and lack of discipline; in other words, low maturity behavior. The appraisal had been planned, the evidence had been pre-pared, and all commitments had been secured from all participants and stakeholders weeks before the start of on-site appraisal activities. In our view, panicking was worse than pointless; it would have been an indicator of a low maturity organization.

This section reveals some techniques you can use to conduct your own informal, relatively inexpensive analysis of the organization's pro-cesses against CMMI practices. These techniques can only be implemented by people who possess a very good, in-depth understanding of CMMI and a deep knowledge of their organization. If you or the organization does not have this knowledge, you should consider hiring external exper-tise to plan and lead the baseline appraisal. Read Chapter 6 — Acquiring Process Expertise and Tools — for more information on hiring and using process and CMMI consultants.

If you already know the current state of your organization in terms of CMMI-based processes, you can scan the rest of this section or skip it and go to "What the Organization Needs or Wants (but Does Not Have)" in this chapter to learn how to determine the other half of the true starting point.

Everyday Appraisals

As pointed out earlier, every day in your job people exchange information, much of which indicates the use of organizational, group, or team processes. Everyday Appraisal is the systematic monitoring of that information stream and, using CMMI knowledge and experience, netting those pieces of information that indicate process instantiation. Everyday Appraisal accomplishes three goals simultaneously with the same effort:

1. It builds a picture (over time) of the organization's state of process implementation.
2. It uncovers and reveals the undefined processes that people are following to perform their work; processes which should be the basis for the defined processes that come later (see "Make the Process What People Do" in Chapter 5 — Five Critical Factors in Successful Process Definition).
3. It contributes to the collection of Practice Implementation Indicators (PII) that the organization will eventually need to conduct an appraisal that is compliant with the *Appraisal Requirements for CMMI* (ARC) and SCAMPI.

Everyday Appraisal uses many of the same skills and techniques that are used in an ARC-compliant appraisal, but is less structured, less intrusive, and is performed continuously by one or a few people over time. When appraising or measuring an organization against CMMI, the mistake people make most often is they start looking at documentation for terms used in the model and they talk to people and ask them questions, again in CMMI terms, about their work. The reason this approach doesn't work is because project managers, software developers, system engineers, configuration specialists, architects, and other people in managerial or technical roles don't naturally work, speak, or write in CMMI vernacular. Nor should they; that is not their job.

Listen, Don't Talk So Much

In the course of normal work, we all have conversations with people every day. One practice I have found personally useful to learning, especially when I was new to an organization, was to engage people in conversations about their roles and their work. In Everyday Appraisal, one of the routine activities is engaging people in casual conversations and doing more asking and listening than talking.

The following dialogue represents a typical information-gathering interview conducted by someone who is trying to measure his organization's implementation of the REQM PA in unnatural terms. (PI = Person in process improvement role.)

PI PERSON: Do you develop an understanding with the requirements providers on the meaning of the requirements?

DEVELOPER: Yes. [She doesn't even understand the question. It's not likely that she knows who the "requirements providers" are or that she even knows what is meant by "requirement," but she instinctively figures out that the desired answer is "yes."]

PI PERSON: (not much wiser) How is bidirectional traceability maintained between the requirements and the project plans and work products?

DEVELOPER: We use our configuration management tool for traceability and we write software ... we don't have any work products. [She really has no idea what she is being asked, but she's not about to show that she doesn't understand.]

PI PERSON: (clueless) How do you obtain commitment to the requirements from the relevant stakeholders?

DEVELOPER: Oh, we're totally committed to developing the highest quality software. Uhm ... I'm late for a meeting and got to run. Bye.

It's not too difficult to see what went wrong in the above conversation. It may seem like an exaggeration, but it's unfortunately close to dialogs I have witnessed. The process person, who's obviously somewhat of a novice at his craft, is asking his questions by reading almost verbatim the REQM practices in CMMI. The dialog was a failure from two perspectives:

1. The PI person failed to obtain any useful information about how the organization manages requirements.
2. The PI person most likely alienated the developer. His unnatural, model-based questions probably made her feel inept and she will avoid interaction with the "process geek" in the future. One opportunity is lost and future opportunities are now at risk.

In process work, always try to avoid asking questions in terms that solicit a binary "yes" or "no" response. People are smart; it will take them no time at all to realize that the best answer is almost always "yes." And once you've received a "yes" answer, it becomes more difficult for you to ask "how" because then it is almost as if you don't believe their "yes."

In the second question, the PI person continues his questioning in the language of CMMI. The developer guesses at the meaning of terms such as "bidirectional traceability" and "work products," but her definitions undoubtedly differ from those of the process person. Her pride gets the better of her (as it often does with most of us), so she doesn't admit that she doesn't understand these terms. Instead, she makes assumptions about their meaning and context and answers accordingly. The process person will go away with the impression that the organization is not managing requirements when, in reality, he has no factual information supporting that impression; he didn't ask good questions and he didn't ask them well.

Now, let's take a look at the same interview scenario with the process improvement person trying to get the same information as before, only this time he's asking the questions in a way that will yield a more accurate picture of what's going on in the organization.

PI PERSON: In your own words, tell me how you know what you should be working on … how do you know what it is you're supposed to be coding?

DEVELOPER: Oh, our team leader e-mails us these ACRs — Application Change Requests — that get assigned to our department by the Change Board. The ACR describes what needs to be changed in the product, you know, like adding some type of feature or functionality.

PI PERSON: When you get these ACRs, does anything else change besides the code?

DEVELOPER: Yep … before we can begin working on an ACR, we first have to do an analysis and tell our team lead if the change affects other components or if it affects the original design or architecture of the application. We also have to tell our lead how long it will take to implement the ACR.

PI PERSON: How do you know that it's okay to actually implement the ACR?

DEVELOPER: Well, we have to write up our analysis of the change and the estimate on the ACR form. Our lead then takes that back to the Change Board, which has to approve it before we can start working on it.

PI PERSON: Who is on the Change Board?

DEVELOPER: Oh, lots of people representing different groups like the customers, senior management, and companies we subcontract work to.

Notice how the information gathered by this interview paints a totally different picture of requirements management in the organization than the previous interview? It also uncovers clues about processes in other areas such as CM and PP and the involvement of stakeholders (GP 2.7). The results are different not because the second interview is going after different information, but because the second interview follows a different approach, one that uses language that is more natural to the respondent, the developer.

In the first question, the PI person doesn't make the mistake of assuming that the developer knows anything at all about CMMI, nor does he pose the question to solicit a yes or no answer. Instead, he simply asks an open-ended question to get the developer to talk about how she knows what to work on. The question is asked in a very natural, conversational style. You could even put the person being questioned more at ease by empathetic techniques: "I'm kind of new here and I can't figure out what it is we're supposed to work on; where does that information come from?"

By the second question, our rapidly learning PI person has picked up on the developer's own language by asking the question using the acronym "ACR;" "I'm now one of you because we speak the same language." The PI person also makes the reasonable assumption that the developer uses ACRs to change code and uses that to lead the developer down the path of thinking and talking about what else gets changed by the ACR. A wealth of information comes back from the developer in response, including information related to process areas other than REQM mentioned earlier. Finally, in the third question, the PI person is asking about requirements review and approval, but structures the question in language that is both conversational and nonthreatening.

Two conversations between the same two characters yield dramatically different results all because of the approach, the style, and the language. The second dialog is an enormous success because the PI person comes away with a treasure of information and clues to more information. He has also established a collaborative, trusting relationship and has banked some emotional capital. Does CMM or CMMI teach a person how to do this: how to be a natural conversationalist or how to become a trusted advisor? It does not. So you might put some thought into the inventory of skills possessed by you and other change agents in your organization.

Reviewing Documents with a Natural Eye and an Open Mind

Another way to measure where your organization stands in terms of CMMI is to review documentation. As it is with talking to people, one of the common mistakes made by process people is that they tend to view the software development world through a CMMI-colored lens. Always remember, the model is an abstraction of the real world, not the other way around (unless, of course, you're insane).

In terms of process, there are essentially two types of documents (often called "artifacts"). The first type of artifact is one that provides people with some form of instruction or direction, either at the organizational or project level. Such artifacts are often generically called processes. The second major type of artifact is often called "implementation," "instantiation," or "evidence-of-use." These are the things that get produced when people follow the processes. One analogy that might help you distinguish the two types of artifacts is baking. When you bake a cake, the recipe (process) and the cake pan or mold (template) constitute the process artifacts. The result — the completed cake — is the implementation artifact; it is evidence-of-use that you followed the processes. The recipe and the mold are not unique to any single cake, rather they can be used to produce many cakes. The cakes are the work products; the outputs of baking.

Within the category of implementation or instantiation artifacts, there are two subcategories which are particularly useful in SCAMPI appraisals: direct and indirect evidence. Direct evidence is the direct result or output of performing a process or procedure. For example, if an organization has a process for developing a project management plan, the direct evidence of that process being followed would be a project management plan built and used by a software or systems project. Indirect evidence includes artifacts that are natural byproducts of people performing a process, but not the direct or intended result. In the example of the project management plan, indirect evidence would include items such as meeting minutes or action items that reflect people discussing, building, or using the project's plan.

Now, let's use an example to illustrate an approach to reviewing a naturally occurring document in an organization. The example we will look at is an e-mail note (see Figure 3.3) from a project manager (Adam) to his senior manager (Tricia).

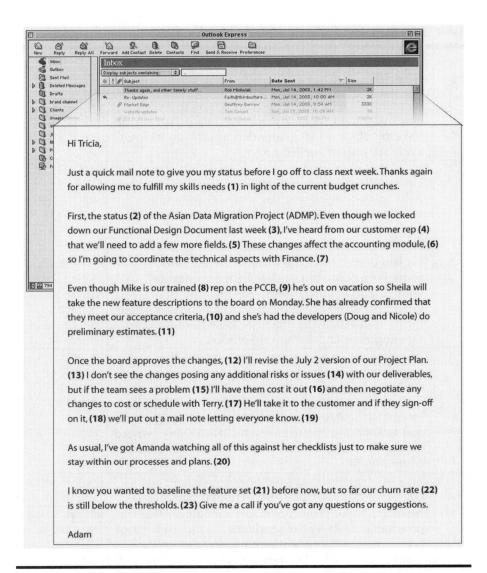

Figure 3.3   Sample E-Mail Note Containing Use of Process Evidence

Notice the numeric tags (in bold type) inserted next to words and phrases in the note. These tags correspond to items listed in Table 3.1, which provides an analysis of how those references relate to CMMI practices.

The e-mail note in Figure 3.3 is, on the surface, ordinary and mundane, not unlike hundreds of e-mail notes, voice mails, and other types of communication that pass between people in any organization every day. However, such routine artifacts can be rich sources of information about

Table 3.1   Mapping a Sample E-Mail Note to CMMI

Tag	Analysis and Clues	Possible Related CMMI PAs and Practices
1	The mention of training needs and going to a class is a clue that training plans and records may exist. Find out the type of class being attended to have an indirect evidence artifact for GP 2.5.	OT: SP 1.1, SP 1.2, SP 2.1, SP 2.2 GP 2.5
2	This e-mail note itself constitutes a project status and indicates that people communicate the status of their work to their managers. An analysis of the full distribution of the e-mail note might lead to an understanding of how relevant stakeholders are kept informed of project status and work.	PMC: SP 1.6 GP 2.10: REQM, PP, PMC, OT
3	"Locked down" could be local vernacular for something being put under change control or configuration management. The mention of a functional design document should be investigated for its applicability to practices in Requirements Development and Technical Solution.	CM: SP 1.2, SP 2.2 RD: SP 3.2 TS: SP 2.1, SP 3.1 GP 2.6: RD, TS
4	The reference to a "customer rep" implies the organization has a sales or marketing unit. This sentence implies coordination with that marketing unit and warrants further investigation for links to Integrated Teaming, Integrated Project Management, and stakeholder involvement. It's also a clue of where to look for more information on how requirements come into the project.	REQM: SP 1.1, SP 1.2, SP 1.3 GP 2.7: REQM, PP, PMC
5	This statement indicates the project manager is cognizant of a change to the requirements.	REQM: SP 1.3
6	The fact that the project manager recognizes that the new fields affect a change to other work products or system functionality (accounting module) indicates that some level of requirements traceability may be performed.	REQM: SP 1.4, SP 1.5 RD: SP 1.1, SP 2.1, SP 2.3 PP: SP 2.6
7	The coordination with Finance provides a clue that there may be native processes or protocols in place that indicate Integrated Teaming, Integrated Project Management, and stakeholder involvement.	PMC: SP 1.2, SP 1.5 GP 2.7: REQM, PP, PMC, RD
8	As with Tag 1, the mention of training needs and going to a class is a clue that training plans and records may exist. In this context, it is reasonable to presume that Mike may be acquiring training for his role on the PCCB (configuration management or change control).	OT: SP 1.1, SP 1.2, SP 2.1, SP 2.2 GP 2.5: CM?

(continued)

Table 3.1 Mapping a Sample E-Mail Note to CMMI (Continued)

Tag	Analysis and Clues	Possible Related CMMI PAs and Practices
9	It would be an informed guess that "PCCB" in the context of this note is an acronym for "Project (or Product) Change Control Board," which indicates a change control or configuration management function is in place.	CM: SP 1.2, SP 1.3, SP 2.1, SP 2.2 GP 2.3: CM GP 2.4: CM
10	Sheila has somehow confirmed that the new feature descriptions meet some "acceptance criteria." The possible existence of acceptance criteria should lead you to look for an organizational standard for requirements, which then should lead you to see if Sheila's confirmation was performed via some type of peer review.	REQM SP 1.1, SP 1.3 RD SP 1.2 VER SP 1.3, SG 2
11	Doug and Nicole have looked at the "new feature descriptions" (read: new or changed requirements) and have developed some type of estimates, presumably to understand the impact of the changes.	REQM SP 1.3 PP SP 1.2, SP 1.4 PMC SP 1.1
12	Another clue of the existence of a change control or configuration management process and resources. If the PCCB comprises cross-functional units and relevant stakeholders, you're also looking at stakeholder involvement (GP 2.7) for both REQM and PMC.	REQM SP 1.2, SP 1.3 CM: SP 1.2, SP 1.3, SP 2.1, SP 2.2 GP 2.3: CM GP 2.4: CM GP 2.7: REQM, PMC
13	Versioning of the project plan indicates a project plan exists and that changes to the plan are managed and controlled.	PP SG 2 PMC SG 1, SG 2 GP 2.6: PP, PMC
14	Project risks are at least being considered. This statement should lead you to look for a project risk management plan and updates to that plan resulting from changes to requirements or project parameters.	PP SP 2.2 PMC SP 1.3 RSKM SG 3 (maybe)
15	Having the team take a look at the changes for a risk analysis indicates there may be a risk management plan and that relevant stakeholders are involved in risk management.	PP SP 2.2 PMC SP 1.3 RSKM SG 3 (maybe) GP 2.7: PMC, RSKM

(continued)

Table 3.1 Mapping a Sample E-Mail Note to CMMI (Continued)

Tag	Analysis and Clues	Possible Related CMMI PAs and Practices
16	Having some team members take a look at the risks and "cost it out" gives us a picture that changes to the project are being somewhat managed and controlled, even if not in accordance with a defined process (GP 3.1).	PP SP 2.2 PMC SP 1.3, SP 1.4, SP 2.1, SP 2.2 RSKM SG 3 (maybe) GP 2.7: PMC, RSKM
17	It would be difficult to imagine that this project could so accurately manage changes to cost and schedule if those two artifacts didn't exist in some form. Given the context of the statement that changes to cost or schedule would be negotiated with Terry should be a strong clue that project commitments were established with relevant stakeholders and that they are monitored.	PP SP 2.1, SP 2.3, SP 2.6, SP 3.3 PMC SP 1.2, SP 1.5, SP 2.2, SP 2.3 GP 2.7: PP, PMC
18, 19	Obtaining customer sign-off on changes to the plan and then communicating those changes are strong indicators of stakeholder involvement in the project.	GP 2.7: REQM, PP, PMC, RSKM
20	Jackpot! This sentence smacks of objective verification of plans and processes. If that beautiful smell doesn't lead you to look into the existence of PPQA practices and processes, I don't know what will.	PPQA GP 2.9: PP, PMC, RSKM
21	Mention of a "baseline" of anything is a clue that configurations have been planned and are being managed. In this case, the product's feature set was apparently baselined and probably so too were other things such as the requirements and the plans.	CM SP 1.1, SP 1.3, SP 2.1, SP 2.2 GP 2.6: REQM
22	"Churn rate" is a term classically associated with measuring the rate of change (volatility) to requirements. This is a good sign that the project (and maybe the organizational unit) has defined and planned measures to quantitatively understand the efficacy of their processes.	MA SP 1.2, SP 1.4, SP 2.1, SP 2.2, SP 2.4 GP 3.2: REQM
23	The observation that Adam knows that the churn rate is still below "acceptable thresholds" might even indicate that, at least in some areas, they are quantitatively managing their processes.	OPP: SP 1.2, SP 1.3, SP 1.4

process implementation in the organization. Even though Adam never mentions a CMMI process area or practice by name, his seemingly innocuous note is virtually dripping with clues to the use of processes (perhaps undocumented) in the organization.

This sample e-mail note is just one example of the treasures of information that float by you every day. Look at the information harvest just from this one note. A 291-word document yields at least 23 indicators or evidence of process implementation in 13 CMMI process areas and dozens of specific and generic practices.

Granted, not all of the hundreds or thousands of pieces of information that cross your desk or desktop every year will be as replete with process clues as our example, but many of them will. So now ask yourself again: are you going to wait until a month or two before your organization's appraisal to start collecting evidence of processes and mapping them to CMMI?

And again: does CMMI teach you how to view a document with an open and curious mind? It does not, yet that is a skill you will need for Everyday Appraisals.

Mapping What You Discover to CMMI

When you capture information from Everyday Appraisals, you will want to record it to begin building the profile of your organization as measured by CMMI. Tracking and recording what the organization does and correlating those process instantiations with CMMI practices is usually called "mapping." There is almost an infinite number of ways to perform CMMI mapping, ranging from using databases and automated document management systems that cost hundreds of thousands of dollars to simple Microsoft® (MS) Word or Excel tables that you can build yourself.

Continual mapping of your organization's process implementation results in two big benefits:

1. You and others in the organization will always know, at any given time, the extent to which you have implemented processes that are consistent with or correlate to CMMI practices. I don't particularly care for the phrase "CMMI compliance," but essentially a current CMMI map will tell you how "compliant" your organization is with CMMI. Thus, some people refer to their CMMI mapping document as a "compliance matrix."
2. When the time comes to begin planning and preparing for an appraisal, you won't have to scramble to find the direct and indirect evidence the appraisal team will need to examine. You will know the names of the artifacts, where they are located, and to which CMMI practices they apply.

Table 3.2 shows a very simple form of CMMI mapping for the Requirements Management (REQM) process area.

What is important to notice is that one artifact or piece of implementation evidence can often apply to more than one practice in more than one process area. This further supports one of this book's premises that organizations should not try to "implement CMMI," which would have people go off and try to create a document for every practice in the model. Instead, organizations should implement process improvements to address their business goals and problems and map those process solutions to the applicable CMMI practices.

As stated, CMMI maps such as the example in Table 3.2 can be very simple, inexpensive, and easy to use. They can also be very complex, expensive, and require a learning curve to become a proficient user. The trade-off between the two ends of that spectrum is usually the utility of the tool and its ability to yield a variety of information for different needs. The MS Word table shown in Table 3.2 is cheap and easy to build and populate, but it is very limited in its ability to yield summarized management information that indicates the organization's overall level of process improvement.

Somewhere between the simple table and the $150,000 database will be the form of CMMI mapping that is right for your organization.

Figure 3.4, Figure 3.5, and Figure 3.6 show examples from a sophisticated MS Excel-based CMMI mapping tool that I developed (Natural SPI CMMI EZTrak) and have used with success in many client organizations. As with most such tools, it has evolved over many years to incorporate experience and lessons learned from its use.

The matrix shown in Figure 3.4 is just a portion of one of many worksheets in the EZTrak tool (there is one worksheet for each PA). As you can see there are columns that provide a count of the number of artifacts (direct and indirect evidence) and the number of organizational process assets. The final column uses a proprietary, tailorable algorithm to determine the extent to which a particular practice is implemented (I), partially implemented (P), or not implemented (N).

The picture in Figure 3.5 shows what is frequently called a "Quilt Chart." Essentially, this worksheet automatically extracts data from other worksheets in the tool to build the grid you see, which provides a view of the organization's implementation of CMMI practices in one view.

The chart shown in Figure 3.6 is automatically generated from the data contained in the Quilt Chart shown in Figure 3.5. This view of CMMI progress and status is very popular with senior and executive managers.

Table 3.2 Example Simple Mapping of Artifacts to CMMI Practices

Goal	Practice	Practice Description	Direct Evidence	Indirect Evidence
SG 1	SP 1.1	Develop an understanding with the requirements providers on the meaning of the requirements.	1. RTM showing original requirements for project Sandstorm and derived requirements 2. Project Bluebook MOA	1. Minutes and action items from 6/26/03 JAD session with customer reps
	SP 1.2	Obtain commitment to the requirements from the project participants.	1. Sales Data Migration project requirements sign-off sheet 2. COLDD conversion SRS	3. Meeting minutes from requirements review with Marketing
	SP1.3	Manage changes to the requirements as they evolve during the project.	1. SDM Requirements Document revision history 2. RMS database change audit report 3. ECR database records	1. CCB meeting minutes and AI log 2. RMS database change audit report and CCB approval records 3. COLDD conversion SRS updates
	SP1.4	Maintain bidirectional traceability among the requirements and the project plans and work products.	1. RMS database (all projects) 2. COLDD RTM 3. IT sustaining engineering work order log	1. CCB meeting minutes and AI log 2. COLDD project review meeting minutes (Requirements section) 3. Project status reports (all projects except COLDD)
	SP1.5	Identify inconsistencies between the project plans and work products and the requirements.	1. SDM requirements peer review defect log 2. COLDD SRS QA audit report	

ID	Goals/Practices	Number of Process Assets	Number of Direct Evidence Items	Number of Indirect Evidence Items	I=Implemented P=Partial N=Not P or I
SG1	Actual performance and progress of the project is monitored against the project plan.				
SP1.1	Monitor the actual values of the project planning parameters against the project plan.	0	3	3	I
SP1.2	Monitor commitments against those identified in the project plan.	4	2	1	P
SP1.3	Monitor risks against those identified in the project plan.	5	4	0	P
SP1.4	Monitor the management of project data.	3	4	0	P
SP1.5	Monitor stakeholder involvement against the project plan.	3	3	0	P
SP1.6	Periodically review the project's progress, performance, and issues.	6	4	3	I
SP1.7	Review the accomplishments and results of the project at selected project milestones.	6	4	2	P
SG2	Corrective actions are managed to closure when the project's performance or results deviate significantly from the plan.				
SP2.1	Collect and analyze the issues and determine the corrective actions necessary to address the issues.	4	3	1	P
SP2.2	Take corrective action on identified issues.	3	2	1	P
SP2.3	Manage corrective actions to closure.	3	2	1	P
GG2	Institutionalize a managed process.				
GP2.1	Establish and organizational policy.	0	0	0	N
GP2.2	Plan the process.	0	3	2	I
GP2.3	Provide resources.	6	1	7	I
GP2.4	Assign responsibility.	0	3	6	I
GP2.5	Train people.	2	1	3	I
GP2.6	Manage configurations.	0	0	0	N
GP2.7	Identify and involve relevant stakeholders.	0	0	0	N
GP2.8	Monitor and control the process				N

Figure 3.4   CMMI Mapping (1 of 3)

A Word of Caution about CMMI Mapping

No one is born with the ability to extrapolate the content of documents and conversations to CMMI process areas and practices. Proficiency in this skill only comes with a very strong understanding of CMMI, extensive experience using this model to implement process improvement in a variety of system engineering and software environments, and reading comprehension skills. Many artifacts will serve as direct evidence for a subset of practices and simultaneously serve as indirect evidence for other practices. Knowing the difference is not intuitive. For the novice, it becomes too easy to look at something and accept on faith that it maps to a component of CMMI. To do so is dangerous in that it can give you a "false positive" with regard to the organization's level of CMMI compliance.

If your organization is relatively new to CMMI, mapping your organization's process implementation to the model may be a skill that is more cost-effective to outsource than to try to grow internally. For more guidance, read Chapter 6 — Acquiring Process Expertise and Tools.

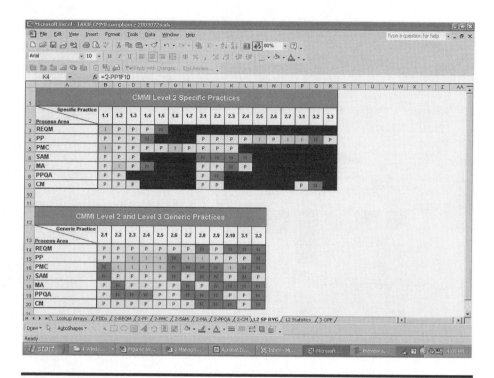

Figure 3.5 CMMI Mapping (2 of 3): Quilt

What the Organization Needs or Wants (but Does Not Have)

Let's assume you do have a pretty good understanding of how the processes and practices in your organization compare with the practices in CMMI. However, the information you probably don't have at your fingertips is what the organization needs or wants, but doesn't have. Here is a bad thought you really want to stay away from: "Let's go do the CMMI practices that we know we're not doing." Why is that a bad thought? How do you know that implementing the CMMI practices the organization is not currently doing would be good for the organization's business? How do you know those practices would benefit anyone in any way? How do you know that implementing those additional process areas or practices won't be a useless waste of time and money?

Here's the good news: it is not solely the responsibility of the people with process responsibility (i.e., the SEPG) to determine the organization's needs and wants. This also happens to be the bad news. Knowing what improvements, process or otherwise, the organization needs to make is based on the organization's goals and strategy. By definition, establishing

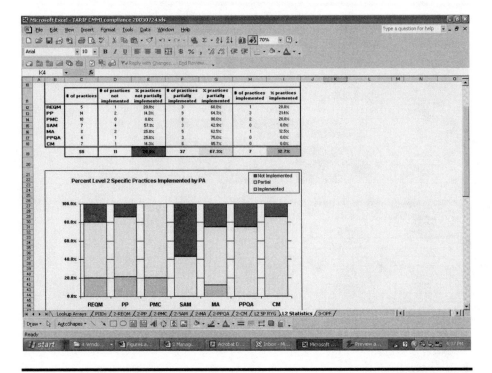

Figure 3.6   CMMI Mapping (3 of 3): Graphs

the organization's goals and strategy must involve the organization's leadership such as executives and senior managers.

Now you have a great opportunity right in front of you! This is a golden opportunity to get the organization's leadership involved in the process improvement effort by asking for their help in understanding the strategy and goals. For more information on the criticality of starting with strategy and goals, read "Everything Starts with the Strategy and It Starts at the Top" in Chapter 1 — News Flash! There Is a Level 1!

GQM Lite: A Quick-Start Approach for Process Improvement

One technique I've used very successfully with clients is a variation of the Goal-Question-Metric (GQM),[19] which Natural SPI has dubbed "GQM Lite." GQM Lite is a relatively low-cost methodology for quickly determining the organization's needs and how those needs can be satisfied with process improvement tasks.

In its most simple form, GQM Lite is a four-step process:

1. Establish the organization's business goals. Goals are statements of desired states in different business perspectives such as customer focus, employee growth and development, financial results, market share, operational excellence, and innovation. The organization's leadership must be personally involved in the establishment of the goals.
2. Define the questions you would need to ask to know if the goals are being met.
3. Define the measures that will answer the questions defined in Step 2.
4. Define the tasks and activities that need to be performed that will yield the measures defined in Step 3.

Establishing the Goals

There are almost an infinite number of ways an organization's leadership can set the strategy and establish the goals for the organization and, with any luck, you won't have to be involved in this task because it is absolutely the hardest thing to do. If the leadership of an organization has not defined or cannot articulate a vision, mission, strategy, or strategic goals, then that leadership doesn't know why its organization exists. And if the organization doesn't have a reason for existing, its potential market and customers surely won't know either. And if the organization doesn't have a vision or mission or strategy or goals, CMMI won't help because process improvement is a means to an end, not the end itself. Leading executives through the job of establishing goals for the organization can be done, but that process warrants its own book and is not discussed herein.

Something that does seem to occur frequently in organizations, especially in government and defense contractor organizations, is that a vision or mission statement will get written but go no further. That's actually good news because you and other stakeholders can at least use a vision or mission statement as a starting point, deconstruct it, and come up with some reasonable business goals.

Such was the case with one client, the Range Instrumentation Technical Support (RITS) organization located at Eglin AFB, Florida. They had a mission statement, which is shown in Figure 3.7. In the figure, key words are circled, which gave us clues as to what was important to the people who wrote the mission statement.

We then led a group of stakeholders to assign meaning and relevance to the four words or phrases that we culled from the mission statement, which led to the following thoughts:

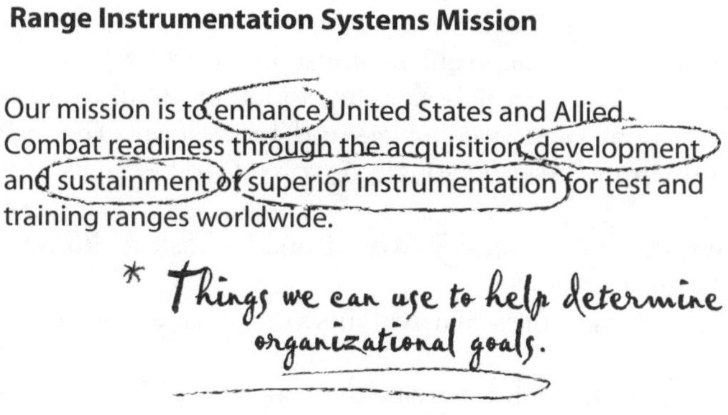

Range Instrumentation Systems Mission

Our mission is to enhance United States and Allied Combat readiness through the acquisition, development, and sustainment of superior instrumentation for test and training ranges worldwide.

*   * *Things we can use to help determine organizational goals.*

Figure 3.7   Mission Statement Deconstruction

- *Enhance:* Improve (make better) capabilities, systems, skills, processes, and quality (also see "Superior Instrumentation" below). We cheated and just looked this one up in a few dictionaries. Reuse.
- *Development:* Enhancement proposals (only?); we left the definition of this term TBD for the time being.
- *Sustainment:* Incremental defect correction and (minor enhancements?).
- *Superior Instrumentation:* We knew what instrumentation means, but what did "superior" mean? We didn't know what it meant to someone else (but then probably neither did they), so we decided that we get to define it!

Focusing on these words and phrases in the mission statement, the client organization was then able to define some long-term business goals, the achievement of which would fulfill the mission.

Getting from Goals to Measures

Okay, so either by miracle or a stroke of fortune or genius, let's say the organization has business goals, and you and the other process improvement people know what those goals are. Now it's time to put some heads together and answer two questions:

1. What questions would we have to ask ourselves in the future (or what indicators would we have to see) that would tell us that the goals are being achieved?
2. What measures would answer the questions that we're asking to find out if the goals are achieved?

To accomplish this step, we have evolved and implemented a one-page form that walks people through the process of deconstructing organizational goals (G) to questions (Q) and then to measures (M). The form, along with a 45-minute tutorial on GQM Lite, enables people to easily define information in these areas:

- What questions should be asked to determine if goals are being achieved?
- Which indicators (measurable concepts) answer the questions?
- What derived measures can be used to form the indicators?
- Which base measures are used to calculate the derived measures?
- Who collects the measures?
- How are the measures collected and how often?
- To whom are the measures reported?
- How are the measures used?
- What methods are used to analyze measures and indicators?

As experience indicates, GQM Lite is both effective and efficient. Following the GQM Lite tutorial, the seven member RITS Engineering Process Group used the GQM Lite form to define all their process measures in a total of about 14 effort hours. I've witnessed process improvement teams or process action teams in other organizations agonize over a measurement program for months and still come up empty-handed.

Figure 3.8 shows one of the interim outputs from using GQM Lite to derive process improvement plans from organizational business goals; it is what success using GQM Lite looks like.

In TYBRIN Corporation, a fast-growing, successful defense contractor, we used GQM Lite with a group of purchasing agents to define the goals, questions, and measures for a revamping of their purchasing and subcontracting processes (SAM and ISM). Figure 3.8 shows the mapping between the procurement goals, questions (measurable concepts), and measures. The measures were then used to determine the tasks and activities for improving the procurement processes.

Getting from Measures to Process Improvement Actions: Managing Process Improvement Requirements

In software and systems delivery projects, the work between customer goals or high level requirements is the analysis, clarification, development, and management of requirements. Requirements management is also the work that needs to occur between establishing the goals and measures and then planning the process improvement tasks.

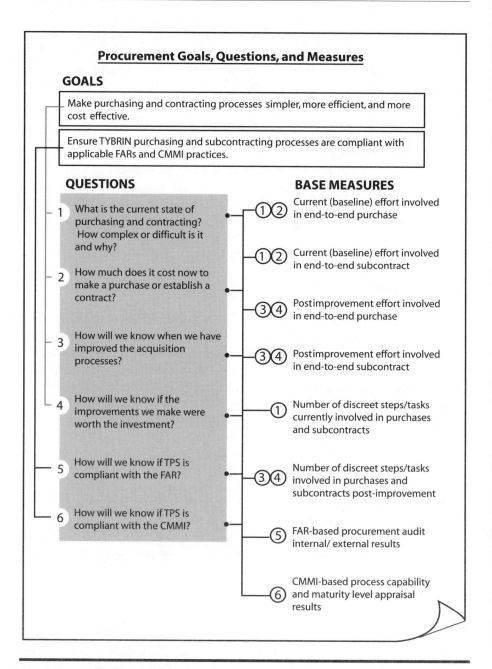

Figure 3.8   TYBRIN Procurement Process Improvement GQM Lite Mapping

There is no reason I can think of why an organization shouldn't gather and document its requirements for undertaking an effort as costly as implementing model-based software process improvement. Having defined the business goals is the first step toward process improvement requirements, but the organization's senior leaders are not the only stakeholders in the process improvement project. For CMMI-based improvements to take hold in the organization and thrive, your project will need the support of many other stakeholders who will want to know what CMMI and process improvement are going to do for them. If you're trying to solve software/system development or integration problems with CMMI, you need to understand those problems because they are very likely to be barriers to the achievement of the goals.

The process for collecting requirements for process improvement is basically the same as collecting requirements for a software or systems project. You talk to the people in your organization — let's call them customers just for discussion purposes — who theoretically are supposed to benefit from process improvement. Ask them what they think they should get in return for investing time and energy into process work. By the way, just because you already know all the standard answers — increased productivity, increased predictability, blah, blah, blah, don't bait or ask people leading questions. Don't ask, "Wouldn't you really like to increase the productivity of your software developers?" Duh. What do you think their answer would be?

In fact, forget about all the cliché reasons for doing process improvement. When you talk to managers, project managers, and engineers and you've convinced them that you really care about their opinion, you'll get some refreshingly creative responses to the question, "What do you want out of process improvement?"

I've heard everything from, "I'm doing it because my manager wants it," (which, by the way, is rarely true), to, "I'm very frustrated with working 40 hours and not knowing where my time went ... I want to know why I can't get more done." As you gather requirements for process improvement, keep in mind that people will often say what they think is expected of them instead of giving you a candid answer. Try not to be naïve. The only real interest is self-interest. Be skeptical of responses to your questions that sound something like "I want to do what benefits the organization." Yeah, right. When you start to get these kinds of answers, modify your approach to get people to trust you enough to tell you what they're really thinking and feeling.

One really good reason for collecting requirements for your process improvement project is that it helps ensure support and buy-in on the improvement activities. If people feel that there's some personal benefit in it for them, they're far more likely to support the initiative. Remember, people do not take action for no reason; everything we do is to either avoid pain or get pleasure or both — process improvement is no different. So find out the problems that plague your organization or find out what would make work easier for people and then try to figure out if a CMMI-based process improvement is going to address any of those things.

However, when all is said and done, here's the very best reason for gathering requirements for process improvement: If you talk to lots of people and no one — not even the leadership — can come up with a single requirement for implementing CMMI, maybe you and your management should ask yourselves, "So, why is it we're doing this?" One of Natural SPI's principals, Donna Voight, would give you this harsh, but sound advice: if you don't have a client, go home. In other words, if your organization's process improvement efforts are not being expended to achieve a business goal or solve a problem, then you don't have a real job. Quit what you're doing and go find one.

In a typical CMMI process improvement project, the people responsible for process definition and implementation will generate lots of documents. As the CMMI project moves forward, you and others will want some assurance that what is being built is satisfying the process improvement requirements and goals.

With your process improvement project's goals and requirements documented, you now know the end-point for your project. You have a deliverable for which you can now do some estimating and planning. If your project is simply process improvement with no clearly defined goals or requirements, you'd better be prepared to be in the job for a long time because you won't know when you're finished. Conversely, you might be in the job for a short time because, at some point, management will grow tired of spending money with no clear deliverables in sight and no measurable progress.

PLANNING THE CMMI PROCESS IMPROVEMENT PROJECT

By now, you have hopefully bought into the idea that CMMI-based process improvement, which is the development and delivery of a process system, can and should be managed as a systems delivery project. So, if CMMI implementation is a project, then there is no logical reason why it cannot be estimated and planned just as you would a software or systems project. This section will give you some basic information on how to go about planning your CMMI process improvement project.

Before you start process improvement project planning, banish from your head any thoughts you may have that your organization is the first and only organization to ever try to implement CMMI-based improvements. Then banish from your head the thought that your organization is so unique that its CMMI implementation will have to be radically different from the approach others have taken. Remember, boldly go where everyone has gone before.

There is plenty of existing historical data from other people's process improvement efforts on which you could base realistic estimates. There are lots of people willing to share with you (this book, for example) their experiences and the lessons they learned in their own process efforts. Go to a SEPG conference and just hang out in the bar. People love to tell their CMM/CMMI war stories and sometimes it gets really old listening to them. But you know what? You really can learn from other people's experiences and you can use those experiences by incorporating lessons learned from others into your organization's process improvement project plan. Ask the war-story bar people, "What would you have done differently?" and then use their ideas to plan your organization's process improvement project. Letting other people and organizations make costly mistakes and then using that information to prevent those same mistakes in your organization is nothing short of business genius and it gives your organization a tremendous competitive advantage.

Another rich source of information on CMM and CMMI implementation is contained in the hundreds of technical reports and other publications located on SEI's Web site. You could spend months reading the historical reports of people who have completed the journey on which you're about to embark and, except for your time, the information is free. This is not unchartered territory; there are plenty of maps and you're wasting your time and your company's time if you don't give them a look.

Finally, acquire the mind set that project planning and replanning is a continuous task and does not occur once. Most of what you initially plan — the estimates, the schedule, the risks, the stakeholder involvement, etc. — will change and the plans need to change accordingly. This doesn't mean you're a bad planner; it means you're a smart project manager. You are allowed to be smarter tomorrow than you are today.

One final caveat: Not all aspects of project planning are covered in this chapter; there are many comprehensive sources on that topic. The areas covered herein tend to be those areas which have historically been overlooked in process improvement projects.

Identifying and Involving Stakeholders

One of the shining improvements of CMMI over SW-CMM is the overt definition of stakeholder involvement (GP 2.7). One of the largest and most commonly occurring risks to any project is that stakeholders, who are not identified and involved early in the project, do eventually get involved, which often results in dramatic changes to the project's requirements, deliverables, commitments, and plans.

Do yourself a favor, explicitly identify the stakeholders for the process improvement project, explicitly define their roles and responsibilities, and obtain each stakeholder's explicit commitment to supporting the defined stakeholder involvement.

The people in charge of planning the JT3 CMMI program knew this and wisely addressed it in the very first draft of the CMMI process improvement project management plan (PMP). In addition to a narrative style document in which they provided detailed explanation of the various roles and responsibilities, they also built a table similar to the partial example shown in Table 3.3. JPEG had already selected "V" life cycle model and they approximately knew the activities that needed to occur in each phase of that life cycle. Based on this, they established stakeholder responsibilities by project phase. The reality is that not all the stakeholders are required to have the same level (intensity) or type of involvement in every phase of the project, so planning that involvement by phase is a really smart idea. The table shows only the first three of six defined project phases, but you get the idea.

Another way to document stakeholder involvement that is slightly more compact than the method shown in Table 3.3 is the three-dimensional information matrix shown in Table 3.4. In this type of stakeholder involvement plan, the specific planned type of involvement for each stakeholder is defined for each project task, deliverable, or decision using an abbreviated description of the stakeholder involvement.

Establishing the Process Improvement Project Team

Traditionally, process focus groups such as SEPGs are the default "project team" for the CMM or CMMI process improvement project, even when those people haven't yet had the epiphany that their work is a "project." However, the classical view of what type of people should comprise the SEPG is exactly what puts the process improvement project at risk. If you think about a systems development and delivery life cycle, you realize that you need different people with different knowledge and skills at different phases of the life cycle. The same is true for developing a delivering process system. It is very unlikely that the same set of people chosen to serve in a process focus function (i.e., the SEPG) will possess all the knowledge and skills the process improvement project will need to succeed.

Taking a project view of process improvement, we can overlay a classical (waterfall-based) systems development life cycle with CMMI-based process improvement phases. We can then identify the knowledge and skills that will be needed by the process improvement project team

Table 3.3 Partial Sample Stakeholder Involvement by CMMI Project Phase

Stakeholder(s)	Phase 1: Customer and User Requirements	Phase 2: System Requirements	Phase 3: Analysis and Design
CMMI Program Management Office (PMO)	• Leads the establishment of PMO and CMMI Program Team • Identifies Program goals and requirements • Establishes current state of process capability or organizational maturity via appraisal • Identifies training goals, requirements, and schedule	• Leads all activities to plan the CMMI program and defines plans and subplans • Leads the definition and communication of program roles, responsibilities, and relationships • Negotiates and acquires program resources • Communicates CMMI program progress and successes to MSC	• Provides leadership for establishing PAWGs • Monitors and reports CMMI program performance against plans • Communicates CMMI program progress and successes to in-scope population • Manages program vendors/suppliers • Establishes CMMI program CCB function • Establishes CMMI program measurements
JT3 Process Engineering Group (JPEG)	• Serves on initial Quick-Look appraisal	• Establishes JPEG charter • Assists PMO in planning the CMMI program • Defines JEEP architecture	• Assists in the establishment of PAWGs; provides representation to PAWGs as negotiated • Verifies PAWG outputs against requirements and JEEP architecture • Integrates PAWG design outputs • Manages and controls process assets as members of the CMMI program CCB

(continued)

Table 3.3 Partial Sample Stakeholder Involvement by CMMI Project Phase (Continued)

Stakeholder(s)	Phase 1: Customer and User Requirements	Phase 2: System Requirements	Phase 3: Analysis and Design
Annex Engineering Process Groups (EPGs)	• Participate in Quick-Look as required	• Establish EPG charters	• Assist in the establishment of PAWGs; provide representation to PAWGs as negotiated
Process Area Working Groups (PAWGs)			• Establish PAWGs • Determine gaps and best practices for Process Areas and design process assets in compliance with process requirements and JEEP • Report design progress/status to CMMI PMO and JPEG
Management Steering Committee (MSC)	• Establishes vision and goals for CMMI program • Helps PMO secure senior management commitment • Works with peer senior managers to establish incentive or motivation for process improvement work and success	• Establishes MSC Charter • Assists the PMO and JPEG in planning the CMMI program • Assists the PMO in acquiring program resources	• Assists PMO and JPEG in staffing PAWGs • Provides representation to JPEG • Provides representation to CMMI program CCB • Resolves issues/risks escalated to MSC from the PMO or JPEG
Process Improvement Center	• Represented in JPEG activities • Supports training efforts	• Represented in JPEG activities	• Represented in JPEG activities • Provides representation to PAWGs as negotiated

(continued)

Table 3.3   Partial Sample Stakeholder Involvement by CMMI Project Phase (Continued)

Stakeholder(s)	Phase 1: Customer and User Requirements	Phase 2: System Requirements	Phase 3: Analysis and Design
Range and Senior Management	• Participates in Quick-Look appraisal as required • Establishes incentive or motivational programs for process improvement work and success • Represents the CMMI program to JT3 customers and builds support	• Reviews CMMI program plans and concurs • Commits resources to CMMI program • Provides feedback on CMMI program via MSC	• Assists MSC, PMO, and JPEGs in staffing PAWGs • Solicits CMMI program status/progress and pays attention to it • Work with MSC to help resolve issues/risks escalated from the CMMI PMO
Projects	• Participate in Quick-Look appraisal as required		• Provide representation to PAWGs
JT3 Customers	• Provide input into CMMI program goals and requirements	• Provide feedback on CMMI program via the range and senior management	• Provide feedback on CMMI program via the range and senior management

at different points in time. This knowledge and skills mapping to process improvement life cycle phases is shown in Table 3.5.

Another factor in establishing the process improvement project team is determining what kind of expertise the organization needs or wants to subcontract versus that which it needs to grow and possess internally. For more information on determining outside consulting expertise needs based on SEI's IDEAL[21] process improvement life cycle model, read Chapter 6 — Acquiring Process Expertise and Tools.

Table 3.4 Simple Three-Dimensional Project Stakeholder Plan

Sample Project Stakeholder Communication, Decision, and Responsibility Matrix

Work Product, Activity, or Decision	PM	Project Team	Project QA	Project CM	Range Manager	Customer	Process Group	Role X
						Project Stakeholders (Roles)		
Requirements	Collect, negotiate, document, approve	Review, input	Review, input	Plan and manage change	FYI	Provide, approve		
Changes to requirements	Receive, approve	FYI	FYI	Manage changes	FYI	Provide, approve		
Project charter	Create, distribute, approve	Review, input	Review, input	FYI		FYI		
PMP	Create, distribute, approve, perform, update	Review, input	Review, input	Manage changes	FYI	Review, approve		
Project status reports	Create	FYI	FYI	FYI	FYI	FYI		
Work product n								
Activity m								
Decision z								

Table 3.5　Skills and Knowledge Required by Life Cycle Phases

System Life Cycle Phase	Process Life Cycle Phase	Knowledge and Skills Needed
Project scoping	Project scoping and baseline appraisal	• Project management expertise • Organizational leadership • Strong knowledge of the organization's business and existing policy and process systems • SCAMPI appraisal knowledge and experience • Negotiation, marketing, sales, and team-building knowledge/experience • Presentation skills
Requirements development	Requirements development	• Project management expertise • Requirements definition knowledge/experience • CMMI knowledge • CMMI implementation experience
Project planning	Project planning	• Project management expertise • Estimating and risk management • Strong writing and document management experience • Negotiation skills • Presentation skills
System design/ architecture	Process design/ architecture	• Process design and engineering knowledge • Information mapping • Technical writing • CMMI knowledge
System development (e.g., software coding, hardware manufacture)	Process definition	• CMMI knowledge • Knowledge of existing organizational structures and processes • Strong writing and document management experience
Unit testing	Process training and piloting	• Training knowledge and experience • Customer relationship management • Statistical analysis • Marketing and sales • Strong presentation and communication skills

(continued)

Table 3.5   Skills and Knowledge Required by Life Cycle Phases (Continued)

System Life Cycle Phase	Process Life Cycle Phase	Knowledge and Skills Needed
Integration	Implementation	• Process definition skills • Document configuration management or versioning control skills/experience (i.e., CCB) • Measurement, analysis, and reporting • Strong communication
System test	Appraisal	• SCAMPI knowledge and experience • Information analysis • Objectivity

Establishing the Project Work Breakdown Structure

Defining a WBS for a process improvement project is not that much different from defining a WBS for the development of a software or integrated system. This is one area in which your organization's process experts either need to also possess strong project management skills or they should recruit an experienced project manager to help out.

Where to Start

As with any project, the process improvement WBS will be shaped in large part by the selected life cycle model. When choosing a life cycle, consider the following:

- The goal(s) of this process improvement project. If this project was initiated to correct an acute business problem, you probably want to see results as soon as possible. For example, if the primary goal of the project is to make estimates more realistic and rebuild your customer's faith in them, you might want to consider rapid prototyping. The WBS should reflect quick cycles with immediate improvement in this one area of project planning and project tracking with heavy emphasis on keeping the stakeholders involved.
- The culture of the organization. If the organization is accustomed to delivering a product every six months, for example, you might want to structure your CMMI project to deliver a subset of improvements every six months. You don't need the busy work of changing the components of the culture that are working. This situation

might lend itself to an iterative spiral development and delivery approach.

- Time and money constraints. Being an imperfect world, money and resources tend to be available for a limited time. If you can determine these constraints, you can make sure you deliver "something" with the available resources. Remember it is better to deliver a slimmed down system that works, than the ultimate unfinished system that dies on the vine from lack of resources and never gets delivered. You are probably looking at a waterfall life cycle with a reduced scope in this example.
- Make sure the life cycle you choose is in the list of approved life cycles for the organization and document why you chose this particular lifecycle in the project plan.

The Process Improvement WBS Content

The phases in a process improvement project are very much like the phases in a software or systems development project. This section describes the minimum phases and tasks or activities therein for almost any CMMI-based process improvement project. Although the information herein is presented in a structure that appears to be a traditional waterfall life-cycle model, realize that there is plenty of opportunity for parallel and iterative work to be performed.

There are at least eight distinct phases to a process improvement project, although your organization can decide to combine the work in two or more of these phases:

Phase 1: Establish the process improvement project and characterize current state
Phase 2: Define and baseline the process improvement requirements
Phase 3: Plan the process improvement project
Phase 4: Design and architecture of the process system and assets
Phase 5: Build the process system and assets
Phase 6: Test (pilot) the process system or its components and fix defects
Phase 7: Communicate, train, and implement the process system
Phase 8: Measure and advertise success in improvements

The following subsections briefly describe the critical tasks and activities for each of these phases. Your project's WBS will differ based on the project's goals, scope and applicability, and constraints levied on the project, such as a target date for achieving a maturity level.

Phase 1: Establish the Process Improvement Project and Characterize the Current State

Establishing the process improvement project is primarily a matter of defining the project's initial purpose (goal), scope, applicability, and high-level success criteria. You will also want to at least get a start on defining the project's stakeholders. Before exiting Phase 1, you should negotiate and acquire at least enough resources and funding to execute project planning in Phase 2. The primary physical output of this phase is a document that defines the information described herein and an allocation of resources to execute project planning.

In Phase 2, the organization will define the requirements for process improvement. However, it will be very difficult to define reasonable process system requirements without some knowledge of the organization's current state in terms of process capability or CMMI maturity. Thus, a task to characterize or assess the organization's current processes and their implementation is frequently a large part of Phase 1. Figure 3.9 shows Phase 1 tasks and activities.

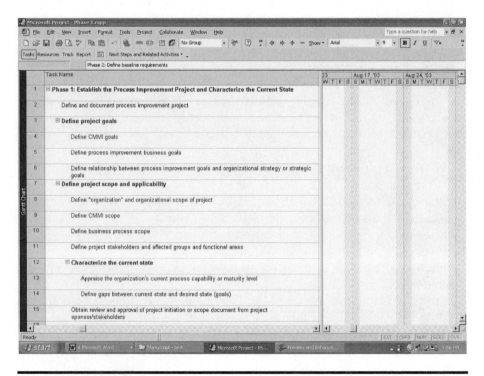

Figure 3.9   Process Improvement Project Phase 1 Tasks

This sample WBS and those shown in the subsequent subsections intentionally do not include scheduling parameters such as start and finish dates and resources. Remember, establishing the WBS is just the first step in establishing the schedule, which cannot be performed until you have more information developed and acquired in the next phase, project planning.

Also, realize and accept the fact that much of what you initially define in this phase can and will change as you gather more information in subsequent phases. When the things defined in Phase 1 do change, it is the process improvement project's responsibility to communicate the changes with the stakeholders.

Phase 2: Define and Baseline the Process Improvement Requirements

As with other projects, the process system to be delivered by the process improvement or CMMI project should be driven by requirements. The process system's requirements are derived from the project's goals, which define the desired state and the current state, which comprises the current process strengths and weaknesses and current business problems that might be resolved through process improvement. If the only requirement anyone can come up with for the process system is that it be used in an appraisal to achieve a CMMI maturity level, you probably don't have a strong enough business case for the project to proceed forward.

The primary output of this phase is a document that defines the baseline requirements for the process improvement project and the process system it will deliver. Again, the stakeholders (as best you can identify them at this phase) should have to review and approve the baseline requirements. As with other projects, the requirements are likely to change in subsequent project phases, so it's a good idea to establish requirements traceability. For more information on defining and managing the process improvement project's requirements, read "Getting from Measures to Process Improvement Actions: Managing Process Improvement Requirements" in this chapter.

Figure 3.10 shows the minimum tasks the WBS should contain for Phase 2.

Phase 3: Plan the Process Improvement Project

Two great sources of guidance for planning a process improvement project are the Project Planning process area of CMMI[2] and PMBOK.[16] Planning the development of a process system includes most of the tasks and activities required to plan a systems project, including:

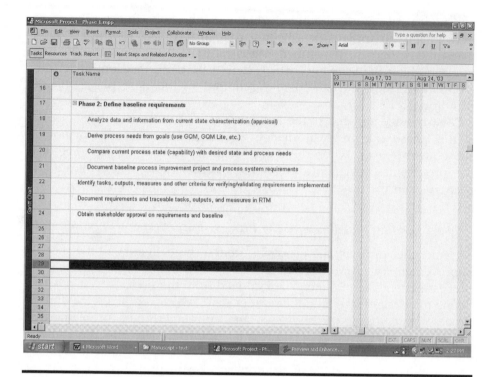

Figure 3.10 Process Improvement Project Phase 2 Tasks

- Establishing the project goals, scope, and stakeholders (Phase 1 in this section)
- Establishing baseline requirements and requirements traceability (Phase 2 in this section)
- Establishing the project team (read "Establishing the Process Improvement Project Team" in this chapter)
- Defining the development or implementation approach and processes
- Defining project assumptions
- Establishing the WBS (this section)
- Estimating cost, effort, and schedule for WBS tasks
- Defining task dependencies, constraints, and critical path or critical chain
- Defining and planning for the management of project risks (look at PP and RSKM)
- Establishing plans for project monitoring, control, reporting, and communication
- Establishing plans for stakeholder involvement and monitoring stakeholder commitments

- Establishing the project and process system configuration management subplan (look at CM and OPD)
- Establishing the project and process system quality assurance subplan (look at PPQA)
- Establishing the plans for testing (piloting, verification, and validation) of the implementation (look at VER and VAL)
- Establishing the process improvement project's measurement subplan (look at MA and OPP)

The primary output of the planning phase is a process improvement project plan which either contains or references all related supporting plans or subplans. Again, the process improvement project team will want to involve relevant stakeholders in the creation of this plan and its review and approval.

Figure 3.11 shows the high-level tasks involved in the process improvement project planning phase.

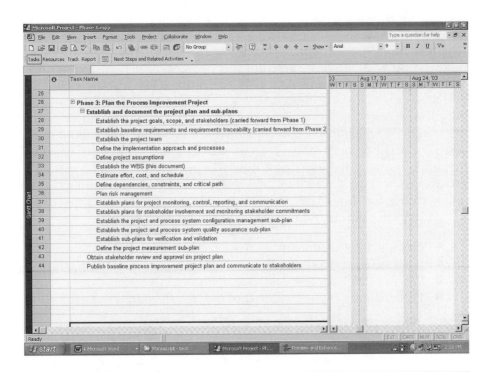

Figure 3.11   Process Improvement Project Phase 3 Tasks

Phase 4: Design and Architecture of the Process System and Assets

The process system, like a software–hardware system, should be built in accordance with an architecture or high-level design and standards. You also need to define the process by which the process system and its components will be designed and built.

The first task in process design is defining the terminology that you want people in the organization to use when talking about different process assets. For example, when someone uses the word "procedure," you want everyone in the organization to have a common understanding of what that term means. For more information on establishing process definitions, read "Define the Process Language for Your Organization" in Chapter 5 — Five Critical Factors in Successful Process Definition.

The next major task in this phase is defining the process by which the process assets will be developed. Will processes and assets be built from scratch or will legacy assets be incrementally improved? What will be the criteria for making such a decision? How and when will process assets be reviewed and what are the processes for accomplishing review and approval? What is the process, method, or mechanism by which the various components of the process system will be integrated or linked together? Will process definition tools be used and, if so, how will they be used and what are the criteria for this decision? When answered, these and other similar questions will yield the process for process definition.

Other critical tasks in this phase include defining the standards — for both format and content — for the various process assets and designing the system by which the process system and its components will be stored and accessed by users. A large factor in designing a process system is first understanding the use cases: how will people use the processes and process assets. Figure 3.12 illustrates the minimum critical high-level tasks and activities for this phase.

The critical outputs from completing this phase are:

- Process asset dictionary
- Process asset standards
- Process definition process description
- Process asset repository design and architecture

(Also read "Design First, then Build" in Chapter 5 — Five Critical Factors in Successful Process Definition.)

Phase 5: Build the Process System and Assets

The fifth phase is to build the processes and assets in accordance with the design and architecture defined in the previous phase and in accordance with the project and process system requirements defined in

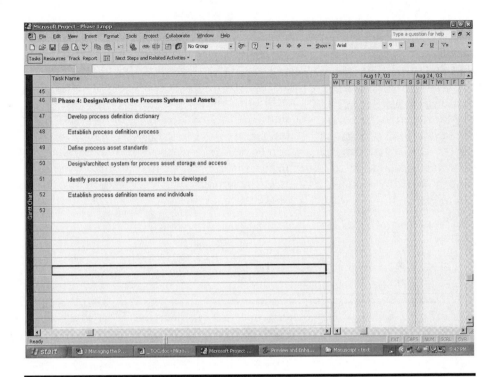

Figure 3.12   Process Improvement Project Phase 4 Tasks

Phase 2. The work in this phase relies heavily on the use of expertise in workflow diagramming and illustration, information mapping, and process/procedural writing. This is not a phase for people new to writing and technical documentation work. Figure 3.13 identifies the critical few tasks in this phase.

Phase 6: Test (Pilot) the Process System or Its Components and Fix Defects

In the vernacular of software or systems engineering, we say we are "testing" the products — the process system in this case — that we have developed. In the language of CMMI, we are verifying the system or validating it.

There are a number of ways to verify that the process system or its components satisfy the requirements and other acceptance criteria such as standards. Peer reviews, walk-throughs, and inspection methods all work, so long as they are performed against objective criteria so that they don't simply become opinion polls.

A powerful lesson I've learned in this phase is that, if you're not careful, you and the process focus or process development people can get caught

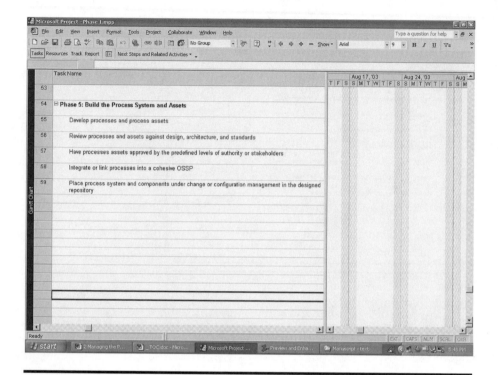

Figure 3.13   Process Improvement Project Phase 5 Tasks

in an almost perpetual loop of review and revision of the process assets. In the first and second reviews, people will catch about 80 percent of the defects. In the third review, they'll catch another 10 to 15 percent of the remaining defects. Now they're addicted to the quest for perfection and overengineering. People will spend 4 to 8 times the effort and money to remove the last 5 percent of defects as they did getting the first 95 percent. This is a miserable return on effort and investment. How do you prevent this? Up front in the planning, establish the exit criteria for verifying the process assets. In other words, get people to agree on a quantitative definition of "good enough." (Read "Process Asset Quality" in Chapter 5 — Five Critical Factors in Successful Process Definition.)

In terms of validation (i.e., the system works as intended in the intended environment), the best way to accomplish this is just put the thing into play. In accordance with the plan for verification and validation, introduce the process system or some components into software/systems projects and let people test drive the stuff. This is where the process team gets tremendous payback on the usability and usefulness of the process assets they have built.

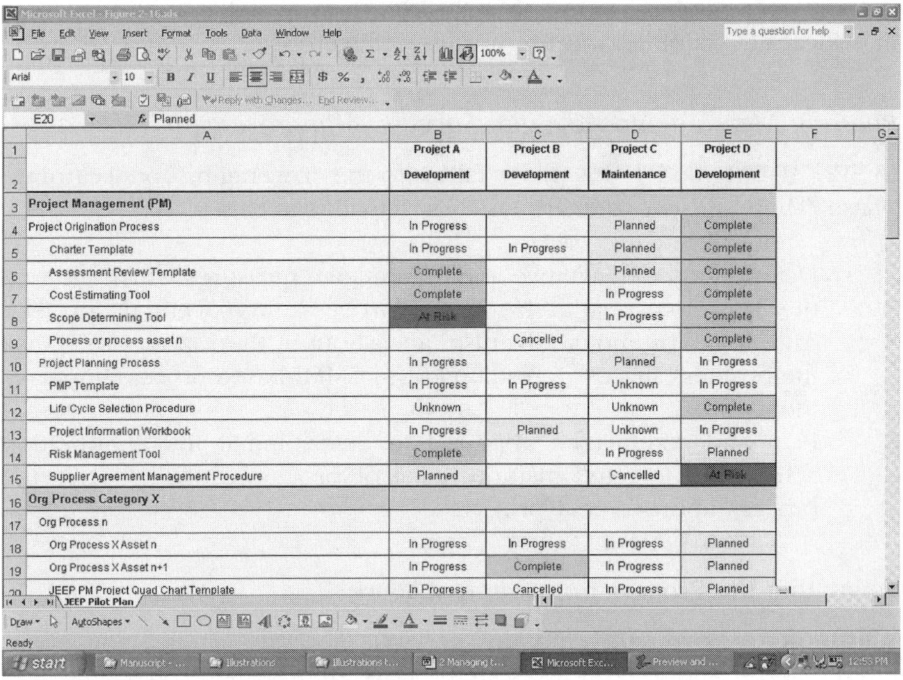

	Project A Development	Project B Development	Project C Maintenance	Project D Development
Project Management (PM)				
Project Origination Process	In Progress		Planned	Complete
Charter Template	In Progress	In Progress	Planned	Complete
Assessment Review Template	Complete		Planned	Complete
Cost Estimating Tool	Complete		In Progress	Complete
Scope Determining Tool	At Risk		In Progress	Complete
Process or process asset n		Cancelled		Complete
Project Planning Process	In Progress		Planned	In Progress
PMP Template	In Progress	In Progress	Unknown	In Progress
Life Cycle Selection Procedure	Unknown		Unknown	Complete
Project Information Workbook	In Progress	Planned	Unknown	In Progress
Risk Management Tool	Complete		In Progress	Planned
Supplier Agreement Management Procedure	Planned		Cancelled	At Risk
Org Process Category X				
Org Process n				
Org Process X Asset n	In Progress	In Progress	In Progress	Planned
Org Process X Asset n+1	In Progress	Complete	In Progress	Planned
JEEP PM Project Quad Chart Template	In Progress	Cancelled	In Progress	Planned

Figure 3.14   JT3 Pilot Plan and Status Matrix

The major word of caution here is: Don't ask people to pilot or validate process assets that are out of phase with the project's current phase or stage. Don't ask project personnel to pilot your project plan template when it is already in system integration and test with the deliverables. There are many other factors that determine which software–systems project pilots which pieces of the process system, but the point is to have a plan. JPEG devised a plan for projects to pilot various components of JEEP that was brilliant in both its comprehension and simplicity. This plan — a matrix — is shown in Figure 3.14. In one single matrix, it provides both the piloting plan and the current status of process piloting efforts.

Phase 7: Communicate, Train, and Implement the Process System

This phase in the process improvement project would be called "release" or "transition" in a software or systems delivery project. The goal of this phase is to transfer the new technology — the processes and process assets — into the community of users and customers. Don't forget that the project needs skill sets that are different from the prior phases to accomplish this phase, primarily people with knowledge and experience in education, training, and knowledge transference.

Phase 8: Measure and Advertise Success in Improvements

In SEI's IDEAL model, this phase falls into the "Leveraging" or "Learning" phase. There are two primary tasks or activities in this phase:

1. Determine the qualitative and quantitative measures which indicate the business results from the process improvement project. Announce, promote, advertise, and shout at the top of your lungs the business successes resulting from CMMI-based process improvement.
2. Plan and conduct an appraisal to determine if the organization achieved a targeted maturity level or process capability from the process improvement project.

Estimating the Process Improvement Work

Whoa there! Have you read the previous sections on defining goals, requirements, and a WBS for your process improvement project? If not, what are you going to estimate? Before you can estimate the work involved in the project, don't you need to first know the tasks and activities by reading the previous section?

Let's say — hypothetically, of course — that your boss has decided that the process improvement project's only goal is to achieve a maturity level and that your boss wants to know how long it will take. How do you respond? Tell him or her to give you one-half hour and you'll be back with a rough guess and then go to do the following …

Visit SEI's Web site at http://www.sei.cmu.edu and conduct a keyword search on "maturity profile." There you'll find the most current version of SEI's maturity profile,[26] which is compiled and published at least twice annually. This profile will inform you how long it takes various kinds of organizations to achieve the maturity levels. Figure out the type of industry and organizational size in which your organization would be categorized in the profile, then look at the short, long, average, and median lengths of time it took others to achieve the maturity level your organization is pursuing. If the number of months make you nervous, add some padding like all project managers do, and give the number to your manager. There, you've just made a schedule estimate using historical data.

Again, planning your process improvement project is an excellent way to "walk the talk" by demonstrating to project managers and others that you're willing to do what you're asking of them. Estimating is just one element of planning; you should also plan and document the project's commitments, assumptions, risks, project team roles and responsibilities, communications, configuration or change management, and process quality assurance.

CMMI Process Improvement Effort and Cost Estimate Considerations

One of the consequences of process improvement efforts not being managed as projects is the fact that there is a lack of historical effort and cost data from such efforts. There is, however, some cost-modeling and scant actual data that you can use to develop a rough order of magnitude (ROM) for your process improvement project.

There are dozens of variables that will affect the cost of your SPI project, but the three variables having the most weight are:

1. What is the starting point? How much work needs to be done to implement processes consistent with the organization's targeted maturity level? Since this is the biggest factor affecting total effort and cost, do not make assumptions about your starting point. Read "Determine the Starting Point for CMMI Process Improvement" in this chapter.
2. How many people are in the organization? Answering this question assumes you and other stakeholders have defined the term "organization." If not, read the section titled "Establish a Common Language" in this chapter.
3. What will be your approach? The approach you choose will dramatically affect the speed and cost of achieving your process improvement goals and the resulting quality of the deliverables. (Read Chapter 4 — Process Improvement Strategies that Work.)

Process Improvement Project Assumptions

As with any project, the success (or failure) of the process improvement project depends on certain assumptions being true and remaining true throughout the life of the project. However, the key to successful project planning is not simply acknowledging this, but getting those assumptions out of peoples' heads, writing them down, and getting consensus on them from the project stakeholders. As in most human endeavors, our egos lead us to assume that everyone else's assumptions are the same as ours. If we proceed to act on this assumption without validating its truth, we put into motion most of the risks our project will incur.

Developing and documenting project assumptions is a difficult endeavor. Even when people can bring their assumptions to the conscious level and articulate them, they often hesitate to do so because they perceive it will reveal too much about the way they view work and their personal agendas. One of the easier techniques for getting stakeholders to reveal to you their assumptions is by starting first. Say to others, "Here are my assumptions about this project; what do you think?" Throw your own assumptions out on the table as "thought starters." Be prepared for people to attack the validity of your assumptions. That's okay, because what you're really after is getting people to open up about their own assumptions. Listen closely when someone tries to disclaim your assumptions because in doing so they may be revealing their own.

After collecting assumptions and getting stakeholder consensus, document them in your project plan. In doing so, make it clear that the assumptions must be true and remain true for the project to succeed. In project planning, the logical opposite of an assumption — that is, an assumption that becomes untrue — is a risk or constraint. In fact, a simple rewording of an assumption can be used in your risk management plan.

Here are some typical assumptions you can modify and reuse in your own process improvement planning:

- Senior management will approve and support the project goals and scope identified in the process improvement project plan in a timely manner (assuming the words "approve" and "support" have been defined).
- Project team members have or will acquire the skills and knowledge needed to perform their tasks.
- Planned and committed resources will remain available to the team.
- The requirements for or scope of this process improvement project will not change significantly once the project is funded and initiated.

Process Improvement Project Risks

 The most important thing to remember about project risks is that they are there whether or not you acknowledge them or do anything about them. Someone smart once said, "the greatest risk is not taking one," which alludes to the opportunities we often find in risks. My corollary to this view is that the greatest risk you can take in a project is the delusion that there are none.

Unlike effort and cost data, there is plenty of publicly available information you can use to plan and manage risks to the organization's CMMI process improvement project. Remember, your organization is not the first to implement CMMI-based process improvements! Consider the following sources.

One of the better sources on the risks to process improvement is from Karl Weigers.[38] Weigers provides ways for you to recognize and mitigate these problems that commonly plague process improvement efforts, including:

- Lack of management commitment
- Unrealistic management expectations
- Time-stingy project leaders
- Inadequate training
- Expecting defined procedures to make people interchangeable
- Failing to scale formal processes to project size
- Process improvement becomes a game
- Process assessments are ineffective

From PMI's PMBOK,[16] we have these ten major reasons for project failure:

1. Inadequate specifications
2. Changing requirements
3. Lack of management change control
4. Inexperienced personnel
5. Unrealistic estimates
6. Subcontractor failure
7. Poor project management
8. Lack of user involvement
9. Expectations not properly set
10. Poor architectural design

With the exception of three of these items — inadequate specifications, subcontractor failure, and poor architectural design — these top causes of failure in software–systems projects can easily also be the top causes of failure in process improvement projects. PMI also lists "Classic Process-Related Mistakes" made in software–systems projects. Following each listed classic mistake below is a brief description of how the mistake is manifested in process improvement projects. The important lesson here is recognizing that the same risks and problems which have historically plagued software and systems projects are most likely the same problems and risks which will plague a process improvement project. Once that is recognized, your organization can employ the same effective risk management techniques used in successful software–systems projects.

Overly Optimistic Schedules

Management sets timeframes for achieving maturity levels without the schedule being supported by historical data or estimates from people actually doing the work.

Insufficient Risk Management

People planning the organization's CMM/CMMI process improvements ignore this section of this book (or any such section in any such book) and just go merrily along practicing another mistake on PMI's list called "wishful thinking." For many, hope will always be the strategy.

Abandonment of Planning Under Pressure

The deadline for achieving the target maturity level is rapidly approaching, so instead of managing the CMMI project in accordance with the plans (assuming there were plans), frenzy sets in. What last week were thought-out, estimated process definition and implementation tasks have now become ad hoc rehearsals for appraisal interviews and retrofitting project documentation with sections intended to satisfy CMMI practices.

Code Like Hell Programming

In process improvement, this equates to "write like hell" process definition. The SEPG and others responsible for process rapidly "roll out" processes by defining all kinds of CMMI-based procedures, putting them in some repository, commanding everyone in the organization to go read them, and then claiming the processes are "implemented" just in time for the appraisal.

MONITORING AND CONTROLLING THE PROCESS IMPROVEMENT PROJECT

This section is short because the lessons are brief. The skills, techniques, methods, and tools involved in managing (aka monitoring and controlling) the process improvement project are indistinguishable from those used to manage software and systems projects. There are at least seven primary concepts to employ in managing the process project:

1. Use the project plans to which stakeholders have committed as the basis for all project performance monitoring and measuring. Let the

committed, documented plans set the expectations, not hallway discussions.

2. Manage the critical, high priority project risks almost continuously.
3. Demand overt concurrence and recommitment to changes to the plan or commitments. Do not accept silence or nodding heads.
4. Continuously market, promote, and build continuing support for the project. Continuously involve the end users of the processes. Put out a periodic newsletter that both informs and entertains people about the process improvement project. Offer small prizes for involvement and interaction.
5. Be decisive and take action when project performance deviates from plans. Do not assume that things will just autocorrect or that leadership will descend from a higher power to help save the project.
6. Always make sure that the process improvement project follows the organization's existing and approved processes for system development and delivery. Never be caught being a hypocrite.
7. Don't be boring. Let's face it, process can be pretty dry stuff, but it doesn't have to be. Learn or find ways to have fun with the project. Reward people, even in small, simple ways, when they exhibit courage and the behaviors that process improvement needs to become institutionalized in the organization.

The Cost, Schedule, and Quality Paradigm

Historically, the project paradigm has been summarized by the cliché, "Good, Fast, or Cheap: Pick any two." I know there's a recent school of thought that suggests you can have all three, but I haven't yet seen that utopian balance attained in process improvement projects.

The problem with process improvement projects, and the one area in which they differ greatly from software and systems projects, is the quality criteria. In a systems project, an organization can establish its own acceptance criteria in terms of product or system defects that are allowable. However, if one of the acceptance criteria for the completion of a CMMI process improvement project is the attainment of a maturity level, then much of the quality criteria are determined by people on the SCAMPI appraisal team who are not members of the appraised organization. In other words, an organization producing a system and an external entity can count the number of defects in a system and come up with the same number. Yet, many organizations go into appraisals thinking they have implemented processes that are fully consistent with the CMMI practices, only to find out the appraisal team sees things differently.

The point is, the organization can't really afford to play with the quality of the process system because the risk is too great (and too expensive!) to target anything less than the highest possible quality of processes and their implementation. This leaves only two variables — cost and schedule — both of which are almost always constrained by people who don't even understand CMMI or the nature and magnitude of the organizational change it represents.

Let's take a truly hypothetical and very nonreal-world view and say that an organization can throw all the money and people in the world at achieving a CMMI maturity level. Couldn't it theoretically move from CMMI Level 1 to CMMI Level 3 in one day? No. The problem is that designing, developing, and implementing a process system has certain sequencing, dependencies, and constraints that logically and physically cannot be bypassed. For example, you cannot measure the performance of a process that does not exist and is not being performed. Thus, even if an organization had a million people on your SEPG and a trillion dollars to spend, they still couldn't implement a meaningful measurements program until some of the processes were defined and implemented.

So now we're down to the last and only variable, the schedule. This is really unfortunate because, as you know, the schedule for CMMI programs are typically established as target dates to achieve a certain maturity level by a certain date. That target is typically not based on any estimates, planning, historical, or industry information or — in some cases — even sanity.

So what do you and the process-focus people do when you realize the process improvement project's performance is not going to meet the targeted date for maturity level attainment?

You do what I had to do once when I was managing a project to move an organization from CMM Level 2 to Level 3 in a targeted time frame of 11 months. You do what any conscientious, responsible project manager does: you raise the issue or risk to those who can make decisions about the project's cost, quality, or schedule and you don't settle for an indecision.

My management in this situation offered to throw more money (i.e., people) at the project. After all, isn't that always the solution? However, I managed to convince them that just more people and not the right skills would be a phenomenal waste of their money. They asked if we could plan less quality into the process system. I told them yes and politely pointed out that they then might spend $40,000 on a CBA IPI and not achieve their desired maturity level. What do you think they asked about next, the schedule?

No. They said, "Well, what if we just get another person to replace you who can do it." I replied, "go for it," knowing full well that in business you can never bluff, because, if you're called on it and you back down, you've lost all your bargaining power from then on. I had to be ready and willing to back up my words with action; I had to be willing to walk. The schedule changed by one month, which is all the more time the project needed.

For more information to understand the dynamics of cost, quality, and schedule as these parameters relate to process improvement, read "Process Improvement: Good, Fast, or Cheap" in Chapter 4 — Process Improvement Strategies that Work.

DO'S AND DON'TS

Now that you've read about the most important aspects of planning and managing a CMMI process improvement project, here are some but not all of the lessons in abbreviated form.

Do

- Reuse as much as possible.
- Make sure everyone is speaking the same language as a prerequisite to starting the CMMI process improvement project.
- Use appraisals or other methods to characterize the organization's current state of process implementation to determine the starting point for process improvement.
- Align the business goals for process improvement (desired state) with the higher level organizational goals or strategy.
- Employ proven project management techniques to planning and managing the organization's CMMI process improvement project because they work.
- Understand that the process improvement project team will need different skills and knowledge at different times in the project's life cycle.
- Recognize and plan for having very little room to maneuver in trading cost, quality, and schedule in a CMMI-based process improvement project.
- In developing and delivering a process system, make sure the project team and stakeholders always exhibit the behaviors they expect to see in others.

Don't

- Don't assume and don't let others assume that your organization must invent its own way to implement CMMI-based processes.
- Don't think that CMMI-based process improvement must be managed differently than software or systems projects.
- Don't spend hundreds of thousands of your organization's dollars if CMMI or process improvement work cannot be correlated with any of the organization's business goals.

WHAT DID YOU LEARN? WHAT WILL YOU DO?

Now take the post-chapter quiz (Figure 3.15), and think about what you've learned and how some of your views toward CMMI-based process improvement have changed. Think about what you will do with the information you've learned (and how it makes you feel).

1. **Which of the following terms are universally understood and have the same meaning for everyone:**
    a. Organization
    b. Project
    c. Requirement
    d. Goal
    e. All of the above
    f. None of the above

2. **True or False:** Every organization's implementation of the CMMI varies; therefore, organizations cannot reuse information or lessons learned from others' process improvement work.

3. **Which of the following statements about managing a CMMI-based process improvement project are false (may be more than one answer):**
    a. You can use a process system project management plan to effectively manage the CMMI work.
    b. There is lots of information and data available on process improvement effort and cost.
    c. Most of the risks and problems faced by systems projects also exist in process improvement projects.
    d. The people who define the process measures for the MA process area should just use their best judgment and experience to come up with those measures.

4. **The most important idea I learned from this chapter is:**

    _____

    _____

5. **I will apply this idea in my process improvement work by:**

    _____

    _____

Figure 3.15   Chapter 3: What Did You Learn? What Will You Do?

4

PROCESS IMPROVEMENT STRATEGIES THAT WORK

We need theories because our brains are small. There are just too many facts and stories to remember without having theories to somehow categorize it all.

— **Stan Rifkin**

WHAT DO YOU THINK? WHAT DO YOU BELIEVE?

Take a minute and answer the questions in Figure 4.1. Then, once you've finished reading this chapter, take the quiz in Figure 4.10 ("What Did You Learn? What Will You Do") to find out how much this information has helped you with your own CMMI-based process improvement. Remember, these quizzes are like process improvement work; there is rarely a right or wrong answer. There are only answers that best suit your organization's business needs.

THE MODEL AND THE REALITY

Here's an interesting thought that is more than academic: If you pick up CMMI and start reading it, if you go attend the "Introduction to CMMI" class, or if you are attending the SEPG Conference for the first time, either you or someone in your organization has already made the decision to use one of the CMMs to improve the organization's process capability or maturity. Whether that decision was made consciously or unconsciously (assumed), whether the decision was based on facts and rationale or not,

1. **From which of the following disciplines can we borrow ideas to use in CMMI-based process improvement:**
   a. Systems engineering
   b. Project management
   c. Organizational learning
   d. Sociology
   e. Systems thinking
   f. All of the above
   g. None of the above

2. **True or False:** If applied correctly, CMMI-based processes will resolve most of the problems typically seen in software or systems development and delivery.

3. **Which of the following statements about managing a CMMI-based process improvement project are true (may be more than one answer):**
   a. Sometimes, if you apply process discipline or new processes to an area of the business, the problems will actually get worse.
   b. The SEI and the CMMI are at fault for making organizations try to achieve maturity levels within ridiculous time frames.
   c. If the organization applies more resources to CMMI and process improvement, it has a much better chance of achieving a maturity level.
   d. The most cost-effective way to implement new procedures is to announce they're in the PAL and then make people go use them. Surprise QA audits ensure implementation.

4. **True or False:** If an organization is just starting out using the CMM or CMMI for the first time, you will probably have to just throw away any legacy processes or procedures they may have in place.

5. **Which of the following two items do you think would get the most results in improving project planning in your organization:**
   a. A very thorough, detailed procedure that tells project managers exactly how to build a project plan.
   b. A project plan template that contains embedded instructions and examples that shows people how to build a project plan.

Figure 4.1   Chapter 4: What Do You Think? What Do You Believe?

whether anyone thinks the decision is right or wrong, the simple irrefutable fact remains that the subsequent actions indicate a decision was made.

Once that decision is made and once the organization starts going down the path of CMMI-based process improvement, the organization will encounter almost nothing in the literature or at conferences that give it reason to question that fundamental decision. Certainly there is nothing in CMMI itself that would give anyone in the organization a reason to challenge or question using CMMI or investing in process improvement at all. And why should it? If someone decided that the organization is going to base its process improvement on a book, it's not the book's obligation to provide counterarguments to that decision.

Most of us operate with the assumption that process improvement is inherently good for the organization. Many of us also accept that CMMI is a good model upon which process discipline and improvement can be based. This chapter does not challenge those assumptions, but it does challenge some of the conventional wisdom underlying some of the currently pervasive approaches to CMMI-based process improvement. The wrong time for an organization to find out that its thinking and decisions about process improvement were misguided is after it has spent millions of dollars. The right time to figure that out is now.

THINKING OUTSIDE CMMI FOR PROCESS IMPROVEMENT

This chapter describes numerous ways of looking at and approaching process improvement and, as such, its sections can be read in any order. It does not prescribe one approach or another, rather it provides you with enough information to make the choices yourself. As with most other endeavors in work, the approach you take should be one that aligns with your organization's goals and prevailing culture. Also, don't be afraid to start down one path, realize it's not working, and choose a different path. The information in this chapter will probably be most useful to people with process responsibility. However, executives, CIOs, senior managers, and the users of processes will also benefit from an understanding of the many different ways CMMI-based process improvement can be achieved. Also, because this chapter assumes that you have an understanding of the use of the word "organization," make sure you read "Define the Process Language for Your Organization" in Chapter 5 — Five Critical Factors in Successful Process Definition.

Once you and others you're working with understand how you're going to plan and manage your process improvement project (Chapter 3), it's time to start implementing improvements. This chapter describes a number of approaches to implementing CMMI-based process improvements and provides some of the positive and negative aspects of each approach.

After reading about the various approaches, discuss them with people who will be involved in or affected by the improvement efforts (i.e., the relevant stakeholders). Determine the approach or combination of approaches that best fits the culture of your organization. Don't be afraid to start down a path and then realize that you need to change direction. That is the essence of process improvement: learning to do things a better way by measuring and analyzing past performance.

The most important thing to remember is to always try to resist the temptation to implement CMMI practices. Use the model as a guide to implement process improvements that effect a positive change in the business.

As people reporting to you embark on process improvement initiatives, demonstrate your commitment to the efforts by actively seeking status and progress on those efforts. If your process improvement initiative is being managed like a software or systems development project (see Chapter 3), you can incorporate the status and progress reporting of that effort into your normal project reviews. Also, expect your SEPG or process people to occasionally ask you for help in resolving issues with projects and suborganizations. They are looking to you for leadership, decisiveness, and clarity. Here's a great opportunity to distinguish yourself as a leader, not just a manager.

Another way you can demonstrate your support for the CMMI process improvement project is to regularly (i.e., during software and systems project reviews) ask your engineering managers or leads how they are engaged in process improvement. If you find out that a project is not involved, find out why. Devise ways to recognize and reward project personnel who actively support the process improvement project.

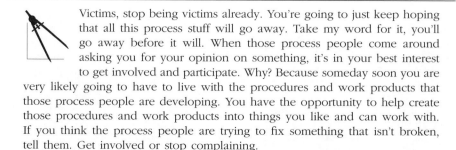

Victims, stop being victims already. You're going to just keep hoping that all this process stuff will go away. Take my word for it, you'll go away before it will. When those process people come around asking you for your opinion on something, it's in your best interest to get involved and participate. Why? Because someday soon you are very likely going to have to live with the procedures and work products that those process people are developing. You have the opportunity to help create those procedures and work products into things you like and can work with. If you think the process people are trying to fix something that isn't broken, tell them. Get involved or stop complaining.

In an organization trying to institute process discipline, you are the people who can make or break the CMMI effort. Project management practices are the crux of most of the CMMI Level 2 process areas and much of what's in Level 3. You are the people for whom change is most critical. Work with your SEPG or improvement teams. Help them understand the development and project management problems you encounter most often. The process people can and will help you find solutions if you let them, and they will address organizational, resource, and process issues that bother you but for which you don't have time to address. Offer to pilot or try out new procedures, procedural changes, and work products such as templates, forms, etc. Doing so gives you an opportunity to give your input into the final products that will be used by the whole organization, thus your chance to make them work the way you want.

APPLYING A SYSTEMS VIEW TO PROCESS IMPROVEMENT

"Systems thinking," which was definitively described by Peter Senge in his seminal work, *The Fifth Discipline*[22] has been used by many people to investigate and resolve deep organizational problems and to achieve higher states of operational excellence. You can use systems thinking to effectively resolve many of the barriers and problems that commonly plague CMM and CMMI-based process improvement initiatives.

Losing Sight of the Forest

Living day to day in a process improvement job, it's easy to quickly lose sight of the big picture; we can't see the forest because we're focused on the trees. We tend to see individuals around us making independent decisions and taking seemingly unrelated actions. We get caught up in trying to deal with every separate event using a different approach and frame of mind than the last event. It is also quite easy to lose sight of the relationships between your process improvement work and product or service delivery. Sometimes, CMMI or process improvement takes on a life of its own and we end up doing process improvement for its own sake.

In systems thinking, Senge described two systems archetypes that provide a way of gaining a big-picture view of commonly occurring systemic problems in organizations: fixes that backfire and shifting the burden. These two archetypes are particularly useful in understanding and resolving problems that frequently plague CMM and CMMI-based process improvement efforts.

Fixes That Backfire

In the fixes that backfire systems archetype, the "obvious" solutions are applied to problems. However, because the perceived or obvious solution is frequently applied hastily and without a thorough understanding of the problem, the result is often unintended consequences, including a worsening of the problem. The most pronounced example of a fix that backfires is corporate downsizing to improve profits. In one 1991 study of 850 companies that had cut staff drastically, only 41 percent had achieved the savings they hoped for.[23]

The diagram in Figure 4.2, known as a "causal loop diagram," illustrates the dynamics of the fixes that backfire archetype as it relates to software and systems process improvement. Because the net, long-term, negative effects of the "fix" are greater than the short-term, positive effects, the reinforcing loop is the prevailing influence in the system.

In Figure 4.2, the top loop represents the organization attempting to address the problem of poor software or systems delivery by implementing

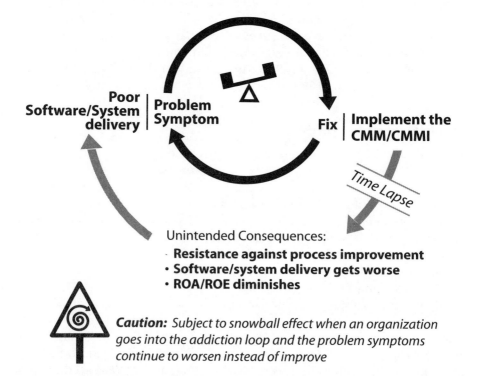

Figure 4.2   Process Improvement — Fixes That Backfire Causal Loop Diagram

CMMI-based process improvement. The bottom loop — also known in systems thinking as an "addiction loop" — represents the long-term effects of the "fixes." When it turns out that process improvement is not the panacea that it is often touted to be, there can be unintended negative consequences.

The fixes that backfire archetype has three primary manifestations in process improvement:

1. The race to achieve a maturity level causes widespread cynicism, which in turn leads to a grass-roots resistance to the process improvement initiative.
2. Process discipline and improvement is done for its own sake and the consequence is that the organization's software and systems development performance gets worse.
3. The cost and value of process improvement activities are not integrated with the price of things sold to a customer, such as the price of products sold or cost of services rendered. Process improvement — and probably other internal "improvement" initiatives — are viewed as infrastructure or overhead. As these internal initiatives grow, so does the percentage of the organization's employees whose work does not directly produce something that is sold to or adds value to what is sold to a customer. Thus, as William Bergquist noted in his book, *The Post Modern Organization,*[24] and as illustrated in Figure 4.3, it becomes increasingly difficult for the expanding organization to achieve and maintain its profit goals because operating costs grow at a faster rate than that of realistically achievable revenue.

The Race to Maturity Levels

Frequently, executive and senior level managers are sold on CMM or CMMI as a vehicle for improving software or systems development and delivery. However, they get fixated on achieving maturity levels under the belief that maturity levels are concrete evidence that processes have improved. The maturity level becomes the evidence that the organization has improved its efficiency, effectiveness, and quality. Such beliefs are as faulty as assuming a student has actually learned something because he received an "A" in class.

Acting on this fixation, the executives will sometimes construct incentives, such as bonus programs for senior and mid-level managers to achieve maturity levels in their respective suborganizations. Now, both level envy and the ensuing race to achieve levels is on! The focus on maturity levels

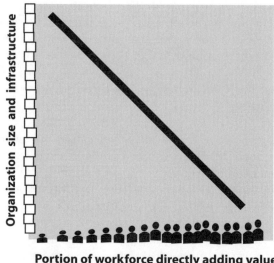

**Portion of workforce directly adding value
to delivered products and/or services**

Figure 4.3 The Relationship between Organization Size, Internal Infrastructure, and Profits

drives people to look for quick fixes that will enable them to "pass" an appraisal. The symptoms — observable behaviors and artifacts — most commonly associated with a race to maturity levels are:

- Deadlines are established for achieving a maturity level without any estimating or project planning that supports the deadline (which, by definition, is a low maturity behavior).
- The appearance of ridiculous slogans such as "Level 5 in '05."
- People read CMM or CMMI and learn to recite the model's terms and phraseology.
- Policies and procedures are rapidly created and frequently copy practices verbatim from CMMI.
- Project managers, under pressure from upper level management, create elaborate project documentation that satisfies the letter but not the intent of the model. They promptly shelve the project documentation and do not use it to manage their projects.

Figure 4.4 illustrates, in a not too exaggerated way, the kinds of behaviors you can often observe in an organization that has entered the maturity level race. You'll laugh, but I've witnessed this behavior and the people involved took themselves seriously!

Figure 4.4   The Race to Maturity Levels

The problem worsens in a large organization in which various suborganizations are trying to achieve maturity levels independent of each other. The suborganizations' respective managers too often let their egos and competitive natures get the better of them and try to out-do each other to be the first to achieve the targeted level.

Once organizations have become entrenched in the level race (or perhaps more accurately, "level wars"), they are in the addiction loop of the fixes that backfire system archetype and you can count on rationale and reason being abandoned. The organization takes a short trajectory to the maturity level, which it more often than not achieves — one way or another.

And what of the unintended consequences? Those incentives for achieving maturity levels usually stop at the mid-level manager and almost never make it down to anyone doing the work, including SEPG members and project managers who have done the bulk of the work in the death-march process improvement project. For a time period ranging from two hours to three days following "passing" the appraisal, everyone in the victorious organization is elated and ecstatic. Once the appraisal high wears off, people start looking around for some lasting and meaningful results of their work but, for reasons they can't always comprehend, everything looks and feels the same as it did before the maturity level race. Now, good luck to executives and senior managers who try to persuade these same people to get excited again about the next maturity level.

Even in extreme command-and-control work environments where the primary motivation is fear (usually of losing your job or being marginalized), these fixes that backfire eventually take their toll on the morale and momentum of process improvement initiatives. Smart, skilled, process-disciplined people start to look outside the organization for more meaningful, rewarding work. Project managers and engineers become jaded on the whole idea of model-based process improvement. They'll continue giving a gratuitous "salute" or lip-service out of fear of not conforming, but they'll be burned out on process. Worse yet, they will perceive — perhaps accurately — that the whole CMMI effort is a waste of time and money. Cynicism prevails.

The organization that chases maturity levels for their own sake and fails to set business goals for process improvement will spend hundreds of thousands or even millions of dollars on model-based process improvement and have nothing more than a few gratified egos to show for a ROI. Recent very public situations tell us that greed and ego may not be so in vogue anymore in corporate America.

Process for Its Own Sake

Sometimes, organizations don't get too wrapped up in CMMI maturity levels, yet process still becomes the be-all and end-all to fixing every issue plaguing the organization. For many of my years in Xerox, "it's the process" or "fix the process" became the only politically acceptable approach to any problem. The primary fallacy of this approach is that it ignores the observable, measurable fact that there really are people and accountability problems which cannot be resolved by addressing only the process.

A classic example of a backfire from applying a fix occurs when organizations attempt to apply the entire CMMI to traditional IT or sustaining engineering shops. With intelligent interpretation and tailoring, many of the CMMI practices can be applied to improve work in these environments. But the implementation of processes and procedures that are nothing but a regurgitation of CMMI in these environments results in burdening the organization with process overhead that doesn't add value, thus making the organization less effective and efficient than they were without CMMI.

Strategies for Fixes That Backfire

Here are some strategies you can employ to prevent or mitigate the effects of implementing fixes that backfire:

- In planning the process improvement effort (see Chapter 3 — Managing the Process Improvement Project), ensure the plans include achieving measurable or observable business goals in addition to achieving maturity levels. Make sure that reporting progress or success includes status against all the process improvement goals and not just the number or percentage of practices satisfied.
- Increase awareness, especially among senior and executive managers, of the unintended consequences of chasing maturity levels. In 2001, I was involved in an e-mail conversation with senior level managers and marketing people in CSC who were trying to come up with strategies for countering the "CMM level race" approach of one of their biggest competitors. I sent out a note thinking that it would be career limiting because it went directly against the prevailing beliefs. Much to my surprise, a senior marketer read and understood my note and invited me into further conversations on how we could market real process improvement benefits without getting into level-wars with our competition.
- Spend the time understanding the problem. You don't necessarily have to conduct time-consuming formal root cause analysis, which can often lead to "analysis paralysis," but you can't afford to continue applying solutions to symptoms, only to have the root problem perpetuate or worsen.
- Establish alliances or relationships with people in your marketing and sales organizations. Find ways to defray some of the cost of process improvement activities by selling (and charging for and collecting!) the benefits of process improvement. With the right approach, a customer or client will be willing to pay a higher price for the goods or services so long as you've convinced them that there is greater value.

Shifting the Burden

According to Senge's, Systems Thinking, shifting the burden ...

> usually begins with a problem *symptom* that prompts someone to intervene and "solve" it. The "solutions" are obvious and immediate; they relieve the problem symptom quickly. But they divert attention away from the real or fundamental source of the problem, which becomes weaker as less attention is paid to it ...

Shifting the burden to process improvement is illustrated in Figure 4.5.

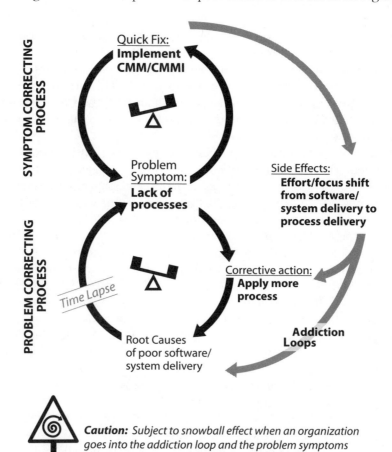

Figure 4.5   Process Improvement — Shifting the Burden Causal Loop Diagram

In this systems archetype, the perceived problem is a lack of process or process discipline in the organization or that the organization does not have a CMM or CMMI maturity level. So the "obvious" solution is to implement CMM or CMMI. In the short term, the fix does appear to address the perceived problem; lack of process is resolved by writing and implementing processes.

The not so easily observed yet insidious effect is that effort and focus shifts from product or service delivery (or integration) to the symptom of inadequate processes, which in turn either does nothing to improve product or service delivery or hurts it by burdening the existing delivery processes with overly bureaucratic standards, procedures, methods, and work products which don't help.

Table 4.1 identifies some common systems development and delivery problems (in the left column). The center column identifies how these problems often are perceived as process (or lack of process) problems. The third column identifies other possible root causes of the problem, which may have little to with process discipline or maturity levels.

What are the unintended consequences of shifting the burden to process? In many cases, doing so can have a compounding, double-negative impact on the organization. With resources diverted from the real problems, the real problems get obscured, are given a lower priority, or are ignored which diminishes the chance of them being resolved. Worse yet, the process improvement efforts — which can be quite expensive — may not only have no discernable effect on the symptoms, they may exacerbate the root cause.

Take the situation in which the outsourced contractor has a poor relationship with the customer or prime contractor. Shifting the burden to process by throwing CMMI, ISO, or Six Sigma or some other initiative at the perceived problem can irritate the customer even more, further worsening the relationship.

Strategies for Dealing with Shifting the Burden

If you are fortunate enough to get involved in a process improvement initiative at its inception, make every effort to get everyone involved in discussing the business problems they want to address and how the perceived process solutions will address those problems. As with fixes that backfire, spend the time first understanding the problem. If the correlation (or better yet, the causation) between business problems and process solutions cannot be clearly established, encourage people to consider pursuing alternative solutions.

If you get involved in a process improvement initiative after it is already underway, be persistent in questioning people around you about their

Table 4.1 Common Shifting the Burden to Process

Software/Systems Delivery Problems	Perceived as a Process Problem	Potential Real or Root Problems Not Addressed Due to Shift
Projects experience scope creep	• Inadequate requirements management process	• Poor business relationship with customer • Lack of discipline in engineering staff • No release strategy • No strategy/process or insertion of new technology • Lack of standards for acceptance or decline of original requirements • The organization doesn't understand the market it is in
Projects overrun cost and schedule	• No estimating or planning process • Project plans are not adequately documented	• Culture encourages low bid; accurate bidders don't get work • "Schedule" is synonymous with "plan" • Poor business relationship with customer • No release strategy • Staff has inadequate skills to do the work • The term "project" is not defined • Management doesn't perceive planning activities as work/progress
Product quality (e.g., defect density) is poor	• No quality process • No people assigned to inspect/audit the quality	• The organization has no defined standards or criteria that define "quality" • Management and the culture rewards fast and cheap; good is not encouraged or rewarded • Staff has inadequate skills and resources to produce quality work
No amount of process improvement activity seems to improve the "bottom line"	• People won't buy into the PI initiative • Not enough in-house CMM/CMMI expertise • Clients don't value the process improvement efforts	• Misalignment between the chosen model and the organization's core business and business goals • There are no baseline performance or capability measurements with which improvement could be ascertained. Improvement is anecdotal.

proposed process solutions. Constantly ask questions such as: "Why are you doing this?" or "What problem will this solve?" You will become quite annoying to some people but, after a while, you'll have them asking the same questions at least quietly to themselves if not overtly.

Again, don't presume that your organization is the first ever to try to apply process solutions to its problems or goals. Conduct benchmarking activities with other organizations to find out what has worked and what has not.

Why Systems Thinking?

Modern software or systems organizations are themselves a system of systems. There are people (social systems), tools and technology (technology systems), and policies, processes, and practices (process systems). The three systems — people, processes, and technology — are inextricably interwoven and changing one without considering the interrelationships can cause fixes that backfire, don't resolve the original problems, or inadvertently make the problems worse.

The greatest unintended consequence of applying a CMMI or process solution to a nonprocess problem is too often the vast waste of resources used for the faux fix. If you really want to improve things in your organization, start by improving the process of process improvement. You can save your organization money and aggravation by using this systemic approach.

PROCESS IMPROVEMENT: GOOD, FAST, OR CHEAP

How you go about implementing process improvement in your organization will be determined to a great extent by the prevailing management and cultural attitudes toward good, fast, or cheap. Every day in every software or system organization in the world, people choose between good, fast, or cheap. These choices are so ingrained, so institutionalized as to almost be the DNA of corporate management. Even in the rare situation in which an individual is conscious that she is making this decision, she won't admit it. More often than not, managers publicly deny making such choices by giving lip service to one choice or the other depending on the audience. To the employees and customers, "quality is our top priority" (aka, good). To the middle and senior managers and to marketing, "we'll meet the deadline" (aka, fast). To the executives and board members, "we'll give you the best value, ROA, or ROA" (aka, cheap).

Most traditional approaches to process improvement that I've experienced or observed achieve only one of these goals: fast. Fast maturity levels also seem to be the favored hot product being sold by many process and CMMI consultants.[25] There are essentially six approaches which heavily influence how good, how fast, or how inexpensive your process improvement project will be. These basic approaches are:

1. Leveraging native standards and practices versus starting from zero (read "Natural Process Improvement through Weeding and Nurturing" in this chapter).
2. Aligning process improvement with the organization's business (read "Business and CMMI Alignment" in this chapter).
3. Learning vicariously versus online learning (read "Organizational Learning and Process Improvement" in this chapter).
4. Focusing on business problems and goals versus focusing on the maturity levels.
5. Implementing a work product-based approach versus procedure-based approach (read the section titled the same in this chapter).
6. Pursuing a meta-feature implementation versus PA or practice implementation (read "Integrated versus Vertical Approaches to Process Improvement" in this chapter).

The following subsections will either help you identify the nature of the process improvement approach that may already be underway in your organization or help you decide which approach is best for your organization. The approaches are then described individually in subsequent sections.

Fast and Expensive Maturity Levels; Not Good Process Improvement

Organizations which experience rapid implementation of CMMI-based process improvement, such as the organizations whose maturity level achievements mark the quick end outliers in SEI's Maturity Profile,[26] are usually after one thing and one thing only: maturity levels (see Figure 4.6). Fast, expensive process improvement is prevalent in organizations that perceive maturity levels as a preeminent marketing tool for getting new business. The segments of industry in this crowd include IT outsourcing organizations trying to out-do competitors in the levels race and non–U.S. companies seeking a piece of the U.S. software and systems market. In these environments, the race to achieve maturity levels resembles a full-on death-march project.

The quality of the results, e.g., the business benefits from process improvement, may be good or poor, but it's likely that no one in the organization will know due to the fact that the only thing tracked, measured, and reported is the achievement of maturity levels.

Figure 4.6 Fast and Expensive Maturity Levels; Not Good Process Improvement

Approaches Employed

Organizations racing to a fast, expensive maturity level typically pursue this combination of approaches:

- They sometimes leverage existing standards and practices knowing that it will get them ahead of schedule getting to the maturity level. However, some organizations will perceive the quickest route to be one involving "wiping the slate clean" or taking the slash-and-burn approach described in Chapter 1 — News Flash! There Is a Level 1!
- Process improvement via CMMI may or may not align with the business goals, but organizations on the fast, expensive track to a maturity level will rarely pause to think about this. In cases where CMMI is obviously an inappropriate model for the organization (a personnel department, a COTS reseller, etc.), the effort to force-fit the model to the business can become very expensive due to time spent trying to find creative ways to implement the model's practices or time spent trying to justify why certain process areas or practices are not applicable.
- Online organizational learning is practiced from the perspective that procedures and work products are simply put into production with the faith or assumption (or indifference) that they work. Learning whether things actually work is not given much consideration.
- The focus is on maturity levels, not solving business problems or addressing business goals.

- A procedure-based approach is employed because it is much quicker to write a procedure than it is to actually modify people's behaviors.
- CMMI practices are used as a "checklist" for tracking and reporting "process improvement" progress.

Additional Characteristics

Some other general characteristics of fast, but not good or cheap process improvement include:

- Instead of being called software or systems "process improvement," the initiative in such organizations is often called "the CMMI project" or "our SEI programs."
- People study for appraisals as if they were tests.
- The technical or business aspects of process improvement are not measured; only progress toward the desired maturity level is measured and reported.
- People in the organization talk about "implementing CMMI," not about improving processes.
- Reward and incentive programs are based on the achievement of maturity levels.
- Relatively large sums of money and effort are expended on getting to maturity levels, but ROI is either never mentioned or is not a concern. Even in fiscally hard times, the budget for CMMI work will be preserved and not cut.
- The standards and procedures, not customer or business needs, become the reason people do things and procedures become the reason people check their brains (and their spines) at the front door when they come into work.

The reason these types of process improvement efforts are so expensive is because relatively lots of money and effort are expended on developing standard policies and procedures without consideration of reusing or making incremental, lower cost improvements to existing engineering and management practices. Also, by nature, such efforts coincide with a management culture that can best be described as centralized command and control. In these types of organizations, a one-size-fits-all mentality often pervades and standardization is sought for its own sake. So, at the level of implementation, much effort is expended by project managers and engineers trying to find ways to force-fit the centralized standard procedures to their projects. If the standards and procedures are overly bureaucratic, effort will be wasted creating project documentation that provides no measurable value to the project, the customer, or the company.

The fast and expensive maturity levels problem is worse than anywhere else in the IT and system engineering outsourcing business. In one such organization in which I was a process improvement manager, the people negotiating new contracts committed the applications personnel to an impossible, no-way-out situation. In their game of outdoing the competition's bids for contracts, the marketers wrote into the contract these two contractual (paraphrased) obligations:

> The organization will be certified (sic) at SEI (sic) Level 3 within 24 months after the cut-over (of application support to the out-sourced vendor).

Many pages later in the contract, you find among the contractual commitments:

> At no cost to the client, the vendor will provide the client with detailed estimates and schedules for projects within three days of the approved service request.

What's so wrong with these two contract points? Think about the behaviors that will be driven by the fact that the client cannot be billed for a critical activity of project planning. If project planning, or at least some of it, has to be free, won't the outsource company's project managers have incentive to produce plans as fast and as cheap as possible? Think about this situation in the context of the Project Planning PA and GP 2.2 of CMMI. Let's face it, no one — not even the best project manager — in this kind of environment is going to establish and maintain realistic plans. Why should they? Where's the incentive?

Fast and Cheap Maturity Levels; Not Good Process Improvement

Some organizations also quickly attain CMMI maturity levels without overspending (see Figure 4.7). At the risk of giving you any not-so-bright ideas, the approaches employed and the general characteristics of such approaches are described in this section.

Approaches Employed

Organizations racing to a maturity level sometimes do so relatively inexpensively. The quality of the process improvement may be good or poor, but no one knows because it's not measured.

Figure 4.7   Fast and Cheap Maturity Levels; Not Good Process Improvement

■ Organizations which achieve fast and cheap maturity levels usually save effort and cost by recognizing existing standards and practices and mapping them to the model.

■ CMM or CMMI usually aligns with the business to some degree.

■ Vicarious learning, simulation, or prototyping are preferred to online learning because they cost less and cause less turmoil in the organization.

■ The focus is on maturity levels, not solving business problems or addressing business goals.

■ A procedure-based approach is employed because it is much quicker and cheaper to write a procedure than it is to actually modify people's behaviors.

■ CMMI key practices are used as a "checklist" for tracking and reporting "process improvement" progress.

Additional Characteristics

Organizations that pursue fast and cheap maturity levels, but not good process improvement also exhibit many of the same characteristics as organizations that go for fast, expensive maturity levels. Some additional general characteristics include:

■ Tools, such as commercially available "CMMI compatible" procedures and templates are purchased and implemented, irrespective of whether or not these things meet the organization's business needs.

■ Organization personnel are trained on CMMI. They are taught how to give the "right answer" in an appraisal interview and often rehearse doing so just prior to the appraisal.

- Project management documentation is created to specifically address CMMI key practices. The documentation is created primarily to showcase in an appraisal, but is generally not used to manage software or systems delivery.

Balanced Process Improvement; Fast Enough, Reasonably Priced, and Good Enough

Whereas achieving the best quality at the lowest price in the shortest time is probably just wishful thinking in any business endeavor, the reality is that you don't have to choose the polarized extremes of any of these goals (see Figure 4.8). In model-based systems process improvement, you can achieve a quality that provides measurable benefits to the organization and its customers in a timeframe that is competitive and at a cost that doesn't kill the organization.

Figure 4.8   Balanced Process Improvement: Fast Enough, Reasonably Priced, and Good Enough

Approaches Employed

The combinations of strategies to take to achieve balanced process improvement are:

- Accelerate your process improvement and CMMI maturity level goals by capitalizing on the project management and engineering practices and standards which have naturally evolved in the organization. Don't spend time and money to create things that may already exist and simply need a little polish.

- Be absolutely certain that CMMI is the appropriate improvement model for your business. (Read "Business and CMMI Alignment" in this chapter.)
- Maximize vicarious learning from others who have gone before you. Make extensive reuse of procedures and work products that have proven successful in similar businesses elsewhere.
- Focus on improving processes to achieve business goals and to solve project management and engineering problems.
- Embed procedures into work products which must be delivered as an interim or final product of your system delivery. (Read "Work Product Based Approach to Process Improvement" in this chapter.)
- View implementing CMMI in terms of its common threads or themes, which we'll call "meta-features." (Read "Integrated versus Vertical Approaches to Process Improvement" in this chapter.)

Characteristics

In organizations which strive to achieve measurable business benefits using CMMI for process improvement, you will often see these characteristics:

- People doing work such as project managers and engineers voluntarily participate in the improvement efforts. They are happy to test out new processes and work products and give the process focus people valuable improvement feedback.
- Individuals bring to the process people practices that they've been doing for some time which they think might help others. They are recognized or rewarded for such behavior.
- Management doesn't stand on your head every other day asking, "When are we going to get to Level n?"
- The organization's performance shows measurable business improvement in measures or indicators such as quality improvement, cycle time reduction, productivity, return on assets, customer satisfaction, employee satisfaction, and innovation.
- When people leave and go to other organizations, they carry with them a positive message about process improvement; they have become apostles of process improvement.
- The desired maturity level is achieved in a timeframe that compares favorably with industry averages.

NATURAL PROCESS IMPROVEMENT THROUGH WEEDING AND NURTURING

What exactly is "natural process improvement" and what is the "slash-and-burn" approach to process improvement? This information in this section assumes you understand the characteristics and consequences of the slash-and-burn approach to process improvement. If not, familiarize yourself with these concepts by reading the applicable text in Chapter 1 — News Flash! There Is a Level 1! This section is devoted to providing you with some concrete examples of the weeding and nurturing approach to process improvement.

In its simplest construct, weeding and nurturing your organization's process improvement effort is exactly what it sounds like. It means temporarily parking and shutting down your bulldozer (CMMI) and taking a look at your organization's processes. It may look like a weed-choked, tangled mess — in other words, chaos — but there are things worth saving in that tangle which you were about to wipe out with your CMMI bulldozer.

Weeding

The first part of weeding and nurturing process improvement is weeding. Weeding an organization's processes is a matter of finding the policies, standards, processes, and procedures which don't help people in the organization do their work and removing those things.

Process weeds are usually not hard to find and you and your process group will find that there is no shortage of volunteers who will point them out. If the process weeds are not obvious, here are some good clues to finding them.

Possible weeds which may need to get yanked out of the organization are those practices people bemoan about. Go investigate the practices about which you hear comments like:

- I don't know why we do this.
- Doing that is a waste of time.
- What do we produce that for … nobody looks at it.
- We haven't done that since 1990.

If people don't volunteer such information, just walk around a bit and start asking questions. Don't interview people. Gain their confidence and trust first and then ask questions like:

- Are there any rules or procedures which you think unnecessarily waste your time or frustrate you?
- What part of your job is the most difficult or frustrating? Why is it frustrating?
- What activities are we doing that our customers complain about?
- What work do you do that takes way more time than you think it should?
- Do you produce anything that you're not particularly proud of or to which you don't pay much attention?

Other possible weeds that are candidates for pulling are documented processes and practices which ostensibly haven't been used by anyone in a long time. If the organization has documented process assets (e.g., policies, processes, procedures, forms, templates, checklists), review them and make note of the date of the last version or revision. If you find process assets that are more than one year old, there's a good chance they're no longer valid. Why? Because over time, people naturally find ways to improve their work. And if they've gone a year or so without updating their defined processes to reflect the way they work, then the defined processes probably no longer affect the work being performed.

An article in the *Harvard Business Review*[36] uses a metaphor similar to our weeds only with a more aquatic theme. The article quotes the Texas insurer USAA; they refer to their process of weeding — simplifying structure and processes — as "painting the bridge."

> That is, once you've finished painting a bridge, prudent maintenance requires that you go back to the other side and start over. So it is with bureaucracy: Once a company has assessed all its core processes and scraped off the bureaucratic barnacles, it's time to begin again.

Be cautious in your search for weeds or your attempt to scrape off the barnacles. Sometimes, you might find a patch of the process garden that looks like it is mostly weeds, yet that patch could belong to someone who will be threatened by you and others wanting to get rid of his weeds. He might not even think that what is growing in his area of the organization is weeds or he might know they are, but it's still *his* patch and you're not allowed to touch it. "Get out of my patch" or "get off my turf" will be the clear message to the process improvement team.

On the other hand, the process people should not assume that just because people don't have a procedure manual open next to them while they're performing work that they're not following defined processes. Truly institutionalized processes are those which have become so ingrained in the culture that people perform them consistently without having to read the documented version.

Finally, remember that what now may look like a process weed to you and others might have at one time been an efficient, well-designed, fruitful process which has now gone to tangle due to neglect. Be aware that some processes and process assets you find might not need to be pulled out of the organization and discarded; they might simply need to be pruned to fit in the current organization.

Nurturing

The second step in weeding and nurturing is nurturing. Process nurturing means finding process assets which look like they might meet the organization's business and process needs if they could just grow a little. In our garden analogy, such process assets and practices just need a little more light, food, and water in the way of design, improvement, and training or communication for them to bear fruit for the organization as an effective and efficient process or practice.

In most software and systems organizations, you will find process "seedlings" that are worthy of your attention. Some of the more common things to look for and nurture are identified in the following subsections.

System or Product Change Practices

No matter how chaotic an organization may appear to be at first glance, I haven't come across one yet that did not have some form of a process or practice for introducing change to its products or services. In the CMMI world, these rudimentary processes can often be nurtured into requirements management processes and assets.

In working with organizations, all of the following terms have been employed to describe some form of a request for a change to an existing product, system, or service:

- Application Report (AR)
- Change Request (CR)
- Engineering Change Request (ECR)
- Enhancement Request (ER)
- Network Service Request
- Service Request (SR)
- Software (or System) Change Request (SCR)
- Software Problem Action Report (SPAR)
- System (or Software) Problem Report (SPR)
- System Enhancement Proposal (SEP)

And I'm sure you know of other similar terms. In all cases, these were instruments or vehicles by which someone was able to request or describe "a condition or capability needed by a user to solve a problem or achieve an objective" or to request or describe "a condition or capability that must be met or possessed by a product or product component to satisfy a contract, standard, specification, etc." — in other words, a requirement.[2] Also, in all instances and implementations in which I've witnessed such instruments being used, there was a process, at least commonly understood and followed even if not defined, for initiating and processing the change instrument.

Such processes and instruments are easy to overlook. People will say, "oh, that's just our bug tracking system" and you instinctively won't think of requirements management although you may be thinking about peer reviews or validation. Be creative and stretch your understanding and interpretation of CMMI and recognize such a process or system as the seedling that it is for a requirements management process or system or a configuration management process.

Project Management

Even in organizations in which it is difficult to uncover any sign of consistent project planning practices, you will still almost always find some consistent project monitoring and control processes. Again, every organization I've worked with has evolved some form of project status or progress reporting or some form of project reviews. Don't throw these practices out with the weeds. Salvage the pieces of these practices that are satisfying a need in the organization and grow them through incremental improvement into project management practices that satisfy the business needs and are consistent with CMMI practices in PMC or IPM.

Quality Assurance

Software and systems delivery organizations have had "quality" departments since the word was first brought into the information industry. Traditionally, the quality organizations have focused on the quality of deliverable products such as hardware, software, and firmware. Yet, think about what the people in quality organizations do, think about the reason for their professional existence. They live and breathe to ensure that something works the way it was supposed to work in the environment in which it is supposed to work. Traditionally, they have achieved their goals through testing. Traditionally, they have come close to satisfying the product "P" in PPQA. It is not a quantum leap to grow quality units into satisfying the process "P" in PPQA.

You can just as easily call a "process" a product. These words are in the public domain and your organization gets to decide what they mean for your organization. As a product, a process can be monitored, tested, checked, looked at, inspected or evaluated to ensure that it also works as intended in the intended environment. Nurture the quality people in the organization to adopt this view and you will soon see PPQA blossom. You'll also be well on your way to establishing some of Verification and most of Validation. Yahoo!

These ideas are just the beginning. Here are some other places to look for promising signs of process life and some clues for nurturing them to maturity:

- *Marketing or sales departments.* (I know what you're thinking, but try to be gracious!) People in sales and marketing, for all their shortcomings, are the life-blood of the organization. They keep you in a paycheck and a mortgage. Yes, they do sometimes sell things that don't yet exist or can't exist, but they also are very much in touch with what the market wants and is willing to pay for. Marketing is the preeminent source for requirements for enhancements to products, systems, and services and, with your help, they can mature into a prime contributor to the feedback loop between the customers and the organization's process improvement work.
- *The help desk or customer support center.* At the back end of the organization (which I'm sure reflects exactly the way people in such organizations often feel) is customer support. This is the other side of requirements; the side that requires a capability in the system product to resolve a problem. Along with marketing and sales, they should be nurtured to be a relevant stakeholder in any REQM and requirements development (RD) process. They will also frequently come up with some amazingly brilliant and simple solutions to problems (think TS).
- *Personnel.* The stigma of personnel departments is tragic. Executives, managers, engineers, and other professionals have for too long viewed personnel departments as a necessary evil. Get over that. In personnel departments, you will often find gold mines of knowledge and skill that you need for your CMMI-based process improvement work. Personnel professionals frequently have a deep understanding of the organization's knowledge and skill baseline and needs (can you say "OT?"). They also possess the psychological, sociological, and anthropological knowledge that the process focus people need to successfully execute an organizational change on the order of CMMI-based process improvement.

- *Accounting and finance.* You'll be surprised at what the "bean counters" count besides dollars (or Euros or Yens or whatevers). Hidden in their formulas and algorithms are all kinds of information and measurements (MA and GP 3.2). Go to them; talk to them. You might be able to get baseline measures for productivity or resource utilization, customer satisfaction, unit production costs (UPC), efficiency, and other measures you will need to later understand the effects of CMMI-based process improvement.

Do you get the picture? CMMI and process improvement is not just for process people and engineers anymore. Almost every team or unit in an organization, almost every individual, has something to contribute to the organization's process improvement work. It is the job of the process experts and change agents to build their CMMI project such that it includes all of these relevant stakeholders and that they are nurtured and fed so that they contribute to the overall success of the organization.

BUSINESS AND CMMI ALIGNMENT

Here is an assertion many readers will not like: Process improvement, and perhaps process capability, is good for every organization; CMMI maturity levels are *not* good for every organization. Blasphemy, you cry! Yes and here's why.

CMMI is a model that provides guidelines for improving software and systems development and delivery, acquisition processes, work management, and process and product integration. It is a model for operational excellence. However, it is not a model for innovation or customer relationship management. Thus, CMMI is not a model for every business in every industry. It is not the grand unification process model, which is as much a fantasy as pixie dust, magic beans, and the "universal business adaptor."[27] Like it or not, CMMI — as with any model — is limited in its application in the real world. This will be unwelcome news to many people who think that CMMI maturity levels are the solution to all their problems.

The lesson is this: Before someone in the organization decides that CMMI is the right model for quality or process improvement or before someone decides to pursue CMMI maturity levels, everyone must first understand the core business of your organization. Let there be no doubt that whatever business you're in, there will be some components (process areas or practices) that your organization will be able to use as guidelines for process improvement. However, adapting some components of CMMI to your business is a far cry from achieving a maturity level.

I remember attending my class at SEI to become a SCAMPI Lead Appraiser. During the course of a class discussion on the model scope for a SCAMPI[SM], there were students who repeatedly asked questions along the lines of "well, what if my organization doesn't do testing, then isn't Validation not applicable?" or "what if we don't subcontract, then isn't SAM not applicable?" C'mon, do you mean to tell me your organization *never* procures any goods or services from an outside source?

The answer to all such questions can be answered by resolving one simple, but very hard question: If your organization doesn't need to use all the process areas of CMMI or, conversely, if all the process areas of CMMI cannot be adopted by your business, why is your organization trying to achieve a maturity level? Why don't you and your leadership simply recognize that only some pieces of the model fit your business and try to improve the organization's process capability in those areas? And at the extreme end of that line of thinking: If there is very little or no relationship between the purpose of the CMMI process areas and practices and the purposes of your business, why is your organization using CMMI at all?

Let's prove the point. Let's take two not-so-hypothetical businesses and explore how well the CMMI process areas fit their business needs. For this exercise, we'll compare a landscaping company with an organizational unit that maintains integrated COTS applications for its customer, a municipal government. For this exercise, we'll only explore the CMMI Level 2 process areas.

Table 4.2 shows the relative translatability and adoptability of the CMMI Level 2 PAs to our two hypothetical businesses. For the "Adoption Difficulty" rating, we use an arbitrary one to five relative scale with a rating of one (1) representing the greatest difficulty in interpreting and adopting the CMMI practices into the business and a rating of five (5) representing the greatest ease of interpretation and adoption. Using this scale, a total score of 35 (seven process areas multiplied by a high score of 5/PA) would indicate a 100 percent fit or ease of adoption or interpretation between an organization's business and the CMMI Level 2 PAs. A total score of 5 would indicate a strong incompatibility between the organization's business and CMMI.

Based on the rationale provided in the table, the landscaping company, with a score of 29, appears to have an easier time interpreting and adopting the CMMI Level 2 PAs than the organization that maintains integrated COTS packages, which gets a score of 18. Surprise! A non-IT organization would have an easier time adopting CMMI practices than would an IT organization!

Maybe the comparison between a landscaping company and a COTS maintenance organization is a bit of a stretch. However, there are plenty

Table 4.2 CMMI Translatability and Adoptability

| Process Area | Landscaping Company | | Integrated COTS Maintenance Company | |
	Adoption Difficulty	CMMI Interpretation	Adoption Difficulty	CMMI Interpretation
REQM	4	Landscaping jobs (projects) differ only in a few requirements such as mowing, trimming, hedging, raking, pruning, etc.	2	It's difficult to determine if one problem or support call equals a requirement or if multiple batched problem reports or calls equal a requirement. People cannot shift their thinking to considering a break-fix as a "requirement."
PP	5	The landscaping job (project is easy to estimate and plan. Size (acreage), effort (person-hours), schedule (total time), and cost (labor + equipment fuel + mileage + maintenance/repair) are all easily quantifiable. Virtually all risks are understood and easily planned.	1	Very difficult to determine what constitutes a project. Too many types of problems and their resolution result in an unusable collection of historical performance on which to base estimates and planning. The standard unit of work performed is difficult to determine.

(continued)

Table 4.2   CMMI Translatability and Adoptability (Continued)

Process Area	Landscaping Company		Integrated COTS Maintenance Company	
	Adoption Difficulty	CMMI Interpretation	Adoption Difficulty	CMMI Interpretation
PMC	5	Very easy to know exactly how much area is mowed or how many feet of hedge trimmed per unit of time. Performance indicators such as earned value are easy to produce and are accurate.	3	Progress and status reporting is easy, but not against vague plans based on the difficulty of defining a "project." An almost infinite source of risks precludes risks being managed.
SAM	3	A limited and finite number of skills required for landscaping makes vendor/source selection and management relatively easy.	2	How does the organization define "supplier" or procurement of the maintenance of COTS. Are the COTS vendors actually vendors or customers or both? Can the COTS integrator get COTS changed? If so, are they a customer or supplier or both.
MA	4	Performance measures are very easy to plan and collect: e.g., linear or square feet mowed or trimmed per unit of time.	3	Ill-defined units of work or units of output make measures almost impossible and mostly meaningless. Some efficiency measures are possible: e.g., number of fixes delivered per unit of time.

(continued)

Table 4.2 CMMI Translatability and Adoptability (Continued)

Process Area	Landscaping Company		Integrated COTS Maintenance Company	
	Adoption Difficulty	CMMI Interpretation	Adoption Difficulty	CMMI Interpretation
PPQA	5	Objective verification of process performance can be performed simply by observing the mowed and trimmed areas to ensure growth was cut at the required level.	2	The organization's processes are determined largely by the constraints of its ability to change COTS products. Floating, ill-defined performance criteria makes objective verification difficult.
CM	3	An almost infinitely variable combination of people, hardware, and processes can accomplish the same results, so CM isn't really important or value adding.	5	Technical limitations of the COTS products and their integration limits the number of configurations to be planned and monitored.

of real-world organizations which will have a very difficult, costly time trying to adopt CMMI practices to their work. Take, for example, an organizational unit which essentially provides integration and governance (oversight) of a delivered system. The organization outsources engineering, project management, and quality assurance. Even though the organization ultimately delivers the system to a customer, how can it be appraised at all the process areas in a CMMI maturity level? Its core business and core competency — outsourcing — only fits some of the model's PAs, primarily SAM and ISM. Sure, the organization could include its subcontractors in the organizational scope of the appraisal, and then you have multiple separate organizations walking away with a maturity level claim from one appraisal, with all of them separately having implemented only parts of the model.

Here's the lesson: You and the leadership of your organization need to ask and find answers to these three questions:

1. What does our organization do? What is our business?
2. What does CMMI do?
3. How well does the answer to Question 1 align with the answer to Question 2?

ORGANIZATIONAL LEARNING AND PROCESS IMPROVEMENT

Before you even consider any undertaking in process improvement, you really should spend some time internalizing and institutionalizing the concepts in this section's title. If you're the kind of person who enjoys developing and building things for the pleasure of saying that you built it, you should seriously consider changing your work habits or finding a different line of work other than process improvement. If you prefer getting results regardless of who did the development, process improvement is for you.

We're all human, we all have some level of ego, and we would all like to believe that the work we do is new, cutting edge, and that we've gone where no one has gone before. Thinking so makes us feel good about ourselves. But the reality is that there really isn't very much new under the sun. The information has always been out there and with the World Wide Web, the information is easy to access. To prove this point, I've often challenged people in presentations and speaking engagements to name a topic on which they believe there is no existing information and to give me 15 minutes on the Web to find some information on the topic. I haven't lost this challenge yet. In process improvement, vicarious learning, reuse, adoption, and adaptation are everything!

What is vicarious learning and why is it so important in process improvement? Put in simple terms, vicarious learning is borrowing existing ideas, processes, approaches, and work products from others and modifying the borrowed things for use in your organization. By reading this book, you are engaged in vicarious learning.

Why are vicarious learning, reuse, and adaptation so important in process improvement? The answer is money. It costs less money to borrow, modify, and reuse things than to build them yourself. There is an even greater cost saving when you borrow things that have already proven successful in other environments similar to yours. Figure 4.9[28] shows the relative cost and accuracy of knowledge or information acquired through the four major approaches to organizational learning.

In Figure 4.9, you can see that the cost of vicarious learning is the least expensive form of organizational learning and that online learning, which is learning by putting an idea or work product into full production, is the most expensive. Following is a discussion of each learning approach in the hierarchy in terms of process improvement and some pros and cons of each.

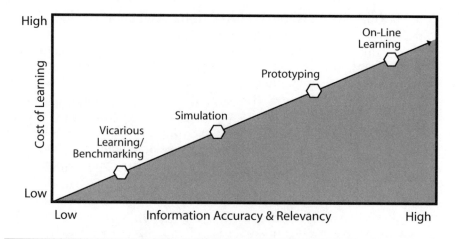

Figure 4.9 The Four Main Types of Organizational Learning

Vicarious Learning and Benchmarking

In terms of process improvement, vicarious learning and benchmarking simply means finding out what other organizations have done in terms of CMM or CMMI implementation and reusing their "best practices" while not repeating their mistakes. The physical activities involved in vicarious learning involve conducting tasks such as:

- Doing Web searches on your topics or questions. One of the richest sources of process improvement information and literature is located on the Software Engineering Institute's Web site: www.sei.cmu.edu.
- Reading books, technical papers, and articles (i.e., what you're doing right now). Including this book, there are more than seven "after market" tomes on CMMI and its implementation.

Example

An example of vicarious learning and benchmarking is finding process-related work products to be used in your organization. Let's say the concept of documenting project plans is new to your organization (yikes!). You and your SEPG or process action team (PAT) can spend weeks discussing and "inventing" a project plan template or you can spend less than 10 minutes downloading one from the IEEE Web site[30] and then modifying it for use in your organization.

The primary physical activities involved in benchmarking are:

- Defining and planning your benchmarking activities and goals
- Searching and reading literature
- Conducting benchmarking visits/interviews with select organizations
- Analyzing the benchmark data collected for use in your organization

In relatively new industries (nanotechnology, for example), the return on effort in vicarious learning may be limited due simply to the relatively small body of knowledge that exists in a truly "new" area. However, this is not the case with software and systems process improvement. The body of knowledge in process improvement is rich in breadth and depth. Thousands of organizations in many different industries, platforms, and geopolitical cultures have used CMM or CMMI to improve their processes.[31] The number of best practices and lessons learned to be found and possibly reused is almost limitless. People whose egos won't let them think that others have gone before them or people who think they're too smart to learn from others will find it easy to say, "yeah, but we're different. We're unique so we can't use other peoples' work." Recognize such statements for what they are: an excuse for being afraid to learn!

A word of caution: Before you run off to start looking for best practices or embarking on a career of benchmarking, make sure you have a working definition for "best practice" and that you have very narrowly defined what you're looking for. Looking at other peoples' work can become so interesting that it's easy to lose sight of your goal. One of the better definitions that I've come across for the phrase "best practice" comes from The Gartner Group:[32]

> A best practice is composed of policies, principles, standards, guidelines, and procedures that contribute to the highest, most resource-effective performance of a discipline. Best practices are based upon a broad range of experience, knowledge, and extensive work with industry leading clients.
>
> There may be no single best practice for any given business process. A process design that works well for experienced, well-trained personnel may be inappropriate for less experienced users. Business processes may assume a prerequisite technology architecture infrastructure or costs that may not be feasible under a different set of circumstances. Globalism may make it unsafe to assume a standard set of factors required to implement successfully a best practice. Therefore, a series of best practices may be defined for each set of circumstances. Gartner Group has recognized that the management of best practices is a "Knowledge Management" challenge.

The key message in the above definition is that even though something is identified as a best practice for another organization, even one that is engaged in a business similar to yours, it is not necessarily a best practice for your organization.

Here are some additional criteria that will help you determine a best practice for your organization.[33] A best practice is one that:

- When implemented will generate or cause measurable improvement in organizational or project performance
- Is widely recognized as a best practice and has a proven track record (be careful of reputation versus demonstrated performance)
- Can be implemented in a wide variety of environments and business uses
- Is consistent with your organization's culture and user patterns
- Is simple and doesn't require five doctorate degrees to comprehend
- Is not dependent on the heroics of individuals to implement or execute
- Can periodically be validated to still be the "best"

Vicarious Learning and Benchmarking Pros

The benefits of vicarious learning and benchmarking are:

- Lower cost relative to other forms of organizational learning (although unplanned or poorly planned benchmarking can get quite expensive).
- Faster speed relative to other forms of learning. For example, you can read the results of five processes used in other organizations much faster than the time it takes to test them in your own organization for results.

Vicarious Learning and Benchmarking Cons

The downside to this approach is that, because no two organizations are identical, the information culled from one organization's process may not be relevant or accurate in your organization. There are many variables affecting the results of process and process improvement methods and tools and not all these variables are always accounted for in published results.

Simulation

Conducting a process improvement simulation is difficult but not impossible. There exists some computer-based process modeling products such as WITNESS[34] and SIMUL8,[35] but most software and systems organizations probably won't spend the initial cost and training costs to purchase and use such packages. A more natural process simulation method is to simply conduct a walk-through or peer review of the proposed process. A few parameters you will want to factor into your process simulation and walk-through are:

■ If there is a current process, make sure you have measurements on its effectiveness and value. If you do not have such data, you will not be able to determine if the new (replacement) or improved process is really an improvement.

■ Make sure you include in the simulation and walk-through people who know the history of the organization. Their experience can provide valuable input in terms of things that have been tried in the past and the results.

■ Make sure you include in the simulation and walk-through people who can or will be involved in the process or affected by it. Obtaining their input on the proposed process helps you get their support for implementing the process once it is ready to be implemented. Their input is also valuable for determining dependencies and constraints within the process.

Example

Let's go back to the situation in which you're trying to implement a project plan template in your organization. Using vicarious learning, you found a template on the Web that you think might work in your organization. Using walk-throughs to simulate the template in use, you (and other process people), stakeholders, and project managers and leads discuss the template and mentally walk it through the overall project planning process to understand how it needs to be modified to have the best chance of working in your organization. At a minimum, the questions that should be answered in these simulations are:

■ Who will use the template and what are their relevant skill levels?
■ When will the template be used?
■ What are its current contents (e.g., sections)? What content exists that is not applicable or not necessary in the organization? What content is it missing?
■ How will people use the template initially and how will they later update the resulting work product?

- What training, instructions, procedures, or other work aids need to be provided in conjunction with the template so that people can use it?
- If the work product (e.g., a project plan) that results from the template being used needs to be approved, how will this be done?
- Which currently existing processes will be affected by the introduction of this template? How will they need to be changed to accommodate the template?

Simulation Pros

The benefits of simulation are:

- Lower cost and faster relative to prototyping and online learning
- It helps garner support for the process, idea, or work product prior to its implementation

Simulation Cons

The downside of simulations for processes, ideas, work products, or practices is:

- No amount of walk-through, peer review, or discussion will ever find all the problems that can be revealed by prototyping or using something in a "production" environment.
- It's very difficult to quantitatively measure or forecast the benefits or results of the new process or practice via simulation.

Prototyping

Second only to vicarious learning and benchmarking, prototyping (or rapid prototyping) is the most cost-effective and beneficial approach for an organization to learn process improvement; consider using the two approaches in conjunction for best results. If prototyping in process improvement were to have a motto, that motto would be "limit the pain." With prototyping, you can find most of the things that are wrong with a process, practice, or work product — things that annoy users or cause them pain — in a small, subpopulation of the organization. You then have the opportunity to remove the problem *before* implementing the new item throughout the whole organization.

Prototyping in process improvement is nothing more than piloting or testing a process, practice, or work product in a limited, controlled environment. As with simulation, before prototyping a process idea, you'll first want to understand as much about the current or existing process as is practical so that you have a "baseline" against which you can compare the new process or work product.

Example

Following the example used in the previous subsections, let's look at how you would prototype the project plan template. The basic steps are:

1. Identify a subpopulation such as a project team to prototype the template. Ideally, the project team will be enthusiastic about trying out a new process solution and providing feedback. The project should also be typical of most of the projects in your organization.
2. Provide any initial training or orientation on the project plan template that is required.
3. Establish an easy-to-use, user-friendly communication channel by which the project personnel can provide usage feedback to the process people or function.
4. Make sure that feedback and suggestions for improvement are encouraged and rewarded or positively recognized. One of the most powerful things you can do in process improvement is to engender a culture in which people in the organization demonstrate their "ownership" of the processes and work products by constantly wanting to participate in their improvement.
5. Collect and analyze the feedback from the prototyping event and incorporate changes into the next version of the project plan template.
6. Based on criteria such as the amount of change or feedback received on the prototyped item, make a decision whether or not it needs to be piloted or tested again before general release.
7. Continue prototyping and revising the item until you and other stakeholders are content that it's ready to introduce into the organization.

Prototyping Pros

The benefits of prototyping are:

- Lower cost and faster than online learning.
- It helps you find significant problems with the proposed process or work product before being released to the general population, thus reducing the cost of resolving the problems. It limits the pain

by having only a few enthusiastic "processnauts" bravely face the problems so that others won't have to.

■ It makes it relatively easy to understand and control the variables and dynamics that affect the success or failure of the process or work product.

■ It helps build support for the eventual process or work product by getting potential users involved in making the thing work the way they want.

 In today's world, corporate executives are all too easily influenced by tool, process, and initiative fads that they read about in a journal or have sold to them by a golfing partner. Without spending too much time understanding the problem, they turn to these "solutions" hoping for a "silver bullet" to fix the organization's problems. Once the tool, process, or initiative becomes the executive's pet project, no one lower in the hierarchy is going to tell him or her that the "solution" is causing more problems than it's fixing. As a change agent in your organization, you will forever win the support, loyalty, and admiration of employees by prototyping or piloting change, thus limiting the pain on the larger organization.

Prototyping Cons

The negatives associated with prototyping processes, ideas, work products, or practices are:

■ Prototyping a process or work product won't be able to find all of its problems. (However, you'll find 80 percent of the problems with 20 percent of the effort.)

■ Prototyping, especially in one iteration or cycle, can yield a "false positive" or a "false negative." The results could lead you to believe that the success (or failure) of the change will be the same when introduced into the larger organization, but there's no guarantee that this is true. You can limit your risk by conducting more than one prototype cycle.

Online Learning

Online learning is taking a change, such as a new or improved process, work product, or tool, and "launching" it; that is, introducing it into real production in the organization. In mature organizations (regardless of the appraised CMMI level), new technology, processes, and work products are not put into production until after they are thoroughly evaluated or tested. Sadly, in this regard, most organizations are not mature. Not enough warnings can be given about the evils of introducing change without the three Ps — planning, preparation, and prototyping — so if you have any say in the matter, just don't let it happen.

WORK PRODUCT-BASED APPROACH TO PROCESS IMPROVEMENT

Over the years, through unscientific studies of numerous organizations taking varied approaches to systems process improvement, through personal involvement in at least eight different efforts to achieve a maturity level, and after hours of surveying the prevailing literature, I have come to the general observation that the vast majority of organizations take what I label a "procedure-based approach" to software and systems process improvement or CMMI implementation.

The procedure-based approach to process improvement works, but it has one major flaw that makes it slow, costly, and painful: The simple, observable, and verifiable truth is that project managers or leads and engineers — people who are doing the work to deliver software and systems — do not like processes and procedures. (I've invested most of my professional life in the process business and even I don't like procedures!)

Why don't managers and technical people like processes and procedures? The answer is found in the understanding that there are essentially two types of technological (including process) change — competency-enhancing and competency-destroying.[37] A competency-enhancing technological change is one which people perceive helps them do their job better, makes them look competent, and makes them feel confident in performing their work. A competency-destroying technological change is one which people perceive inhibits them from doing a good job, makes them look incompetent in an area in which they were previously competent, and then threatens their sense of job security. People resist changes they perceive as competency-destroying for reasons that should be obvious: competency-destroying technology is perceived as a threat to being able to make the mortgage payment or rent.

In the world of CMMI-based process improvement, lengthy, detailed, narrative procedures are a competency-destroying technological change. Why? Because if you give a project manager or engineer a 40-page procedure document and tell them that it describes their job, they're wondering what was wrong with the job they were doing yesterday for which they didn't need a 40-page procedure. It gets worse from there. Let's say you hand a project manager a 30-page procedure telling her how to build a project plan. She reads the instructions, but when she sits down at her PC to build the project plan, she's still staring at a blank screen wondering where and how to start. Prior to the project plan procedure she was able to build project plans and now she can't. The procedure has left her feeling incompetent and threatened and the process people have made an enemy. There is a better way.

In the world of process, things such as electronic or online templates, forms, and checklists are all competency-enhancing process technologies. Why? Because with such items — generically called "job aids" or "implementation assets" — people are able to do the same job they were doing before, only better, faster, and more consistently. A plan template or a procedure checklist in the hands of a manager or technical person gives them an easy way to perform their work without having to remember or go read about things like sequence, format, and style. They're left deciding only on the area that is important, the content. Implementation assets also enable people to do the one thing they believe they're getting paid for, deliver work products. Engineers, architects, analysts, and project managers don't believe they get paid to perform processes; they believe they get paid to produce things. So when the process people give them an easier way to produce things (e.g., an easier way to get paid), it makes them feel more competent and more secure. And the process people have gained a friend and advocate of process improvement.

But wait, you say, what about CMMI GP 3.1 that calls for a "defined process?" Well, if you ensure that your template, form, or checklist is "a documented expression of a set of activities performed to achieve a given purpose that provides an operational definition of the major components of a process," then it is a "process description" even if your organization calls it a template.[2] (For more information on defining your organization's terms for process assets and process definition activities, read "Define the Process Language for Your Organization" in Chapter 5 — Five Critical Factors in Successful Process Definition.)

INTEGRATED VERSUS VERTICAL APPROACHES TO PROCESS IMPROVEMENT

Traditional wisdom and practices have held that the way to "implement CMMI" is to establish improvement teams — commonly given labels such as PATs, Process Improvement Teams (PITS), or PAWGs — to go off and define and implement new organizational processes and procedures along the lines of the process areas in the model, REQM, PP, MA, etc. This approach is perpetuated throughout the CMMI and process improvement community by practitioners and consultants, perhaps because it's too hard or just too scary to think about doing things differently. This vertical or "silo" approach to CMMI-based process improvement usually works, but it's slow and painful and there are better approaches. Traditional wisdom, as it sometimes turns out, can be an oxymoron.

The Downside to a Process Area Approach to CMMI Implementation

Teams charged with focusing process improvement on specific process areas tend to do exactly that — they focus on specific PAs. Experience, my own and others, reveals that this type of approach can and often does result in one or more of the following situations:

- The PA teams develop and define multiple processes which are not linked, integrated, or connected in any way. For example, there is no defined work or data flow from managing requirements to project planning to project monitoring and control. The process definitions are written as if a person or team would or could perform one process area in total isolation from other organizational processes.
- The PA teams collectively develop excessive documentation. There will inevitably be a policy document for managing requirements, another policy document for project planning, and so on. There's nothing in the model that says you can't have just one process policy document that governs all the organization's processes, but this isn't usually what comes out of a vertical approach.
- PA-focused teams tend to overlook or ignore the institutionalization aspects of process improvement. It's too easy to focus on building processes that cover the specific practices and it's also easy to not worry about how the process will be stored, accessed, retrieved, viewed, used, trained, communicated, and updated. Admittedly, CMMI is a vast improvement over SW-CMM in this arena. CMMI provides more explicit guidance for addressing the institutionalization aspects of process with its Generic Goals and Generic Practices.
- PA-focused teams, especially those composed of inexperienced people, will develop processes and process assets mimicking the language of CMMI. When these things are later pushed out to people in the organization (the customers), they are alien and hard to relate to the work and the business. People who have always performed procurement jobs won't know that "Supplier Agreement Management" affects them and is related to their work. "Oh, SAM … that must be some other group of people who do that."

Don't forget: Even though this section talks about process improvement or process action teams, the establishment of such teams may not be necessary, and the organization certainly does not have to hire additional people to staff these teams. On the contrary, the organization will benefit from staffing these teams from existing experienced and knowledgeable personnel.

An Integrated Approach for Developing an Integrated Process System

Remember, CMMI is not prescriptive; it is not a set of instructions which your organization must follow. It is a body of guidelines, which is yours to carve up and implement any way your organization sees fit.

When you've used CMMI to implement process improvement in a variety of environments and enterprises of different type and size, you will start to see some underlying, unheralded, and sometimes unmentioned themes that are integral to implementing CMMI-based processes. But you don't have time to go out and get that experience, so here it is encapsulated for your use.

You already know that CMMI is built from components such as Process Areas, Specific Goals, Specific Practices, Common Features, Generic Goals, etc. But what you may not know is that implementing CMMI-based process improvements involves some meta-components: components which transcend yet are interrelated to all of the CMMI components.

In my consulting firm's work with clients, we have discovered six implementation meta-components; I'm sure there are more and I encourage you to look for them. These six meta-components and why they are important to CMMI implementation are discussed briefly in the following subsections:

1. Standards
2. Communication
3. Traceability and integration
4. Tailoring
5. Document management
6. Process implementation

Standards

No matter what kind of process or process asset the process focus people intend to build, they can build it faster and with higher quality if there is a predefined standard for the item. When most people think of standards as they relate to process documentation, they think of format and style standards, but format and style are almost inconsequential relative to content standards. If your organization hasn't defined standards for the process assets, doing so is going to seem like a hassle and a waste of time. The people doing process definition just want to jump right into writing the processes. They will eventually address the standards issue, but it will cost more later than sooner. For more information on defining standards read "Design the Process Asset First, then Build" in Chapter 5 — Five Critical Factors in Successful Process Definition.

Communication

If a process grows up in an organization and no one hears about it, does it exist? Well, yes, but it's existence is meaningless. Communication is critical to all aspects of the development and implementation of an organization's process system, yet it is not a Process Area or Generic Practice. Look at all the following words, many of which are used in CMMI, that mean, imply, or have their basis in communication:

- Report
- Inform
- Communicate
- Status
- Coordinate
- Review
- Read
- Write
- Talk
- Teach/train
- Present
- Define
- Listen
- Send
- E-mail
- Meet
- Call
- Show
- Describe
- Discuss

So, what if you had a meta-component team that defined what information gets communicated in the organization and how that information gets communicated. Don't you think the other process definition teams could use those standards?

Traceability and Integration

Even if your organization takes the traditional path of setting up improvement teams based on PAs, they can still mitigate many future problems by establishing a team to integrate the disparate processes and process assets and establish and maintain the traceability or linkage between them.

I learned a hard lesson at Xerox. We had PITs which were individually focused on developing processes along the lines of SW-CMM's KPAs. At the end of about 18 months worth of work, these teams delivered some really great processes and process assets. However, before any of this good work could be implemented in the organization, we had to have an unplanned follow-on team define the integration: how the different processes worked together. In retrospect, there's no good reason why we didn't work the integration issues all along except that we just didn't know.

As processes are developed, the process integration team ensures there are interfaces or hand-offs between one process and another and that the traceability between organizational goals, the processes, and the measures collected is defined. This integration work can and should be performed in parallel with process design and definition and should not wait until the processes are completed.

Tailoring and Tailorability

We all recognize that, in theory, the concept of defined, structured process tailoring doesn't kick in until CMM or CMMI Level 3 (in OPD and ISM in SW-CMM and in OPD and IPM in CMMI). However, the reality is that even in organizations striving to achieve no greater maturity level than 2, business demands that projects and suborganizations be able to tailor the organizational processes and process assets. So, your choice boils down to either (1) ignore defined and managed process tailoring until the organization is moving toward Level 3 or (2) recognize that there are significant business benefits to introducing structured process tailoring early in the process improvement journey. As always, you decide.

Here's a way to get ahead of the curve on tailoring. Establish a process definition team that works in parallel with the other process definition teams. The charter of this team — let's call them the Tailors — is to review all the processes, procedures, and process implementation work products being designed and developed and define reasonable tailoring criteria and guidelines based on the needs of the enterprise. As the other process teams develop the processes and assets, the Tailors will be collecting historical data from the projects and using that data to establish fact-based categories or types of projects. The Tailors can then use the factors that determine project types to define realistic tailoring criteria and guidelines, which will be ready to pilot at the same time the processes are ready.

Document Management

Look at all the types of work and activities involved in designing and defining processes and process implementation assets:

- Information gathering and organizing
- Writing
- Learning or training in document production or publication systems and processes
- Editing for content, standards, style, grammar, spelling, etc.
- Reading, reviewing, and providing input
- Building forms
- Establishing documentation links and cross-references
- Establishing and maintaining version control
- Uploading documents to online libraries and maintaining integrity

What all of this adds up to is technical publication and document management, which are not the typical skills people look for when they establish a process focus and process definition team. SEPGs and other process focus groups tend to recruit people with knowledge of CMM and CMMI and people with technical expertise such as project management, configuration management, etc. Yet most of the work such groups perform is the creation of documents (with the word "document" meaning some physical or electronic representation of information).

So, the meta-component is document creation, production, and management and you don't really want to pay highly skilled technical people or CMMI experts to struggle with MS Word doing work which they're not really good at doing. Establish a process subteam to have responsibility and authority to have a strong say in document design and to perform the production editing and formatting.

Process Implementation

Too often, the focus of people with process responsibility is the definition of the process or procedure document and not much thought is given to helping the users of processes implement them. There is a large and often overlooked or underestimated chasm between a process definition and its implementation. (Also read "Work Product Based Approach to Process Improvement" in this chapter.) There is also a difference between the skills required to define processes and the skills required to read a process and build things that can be used to implement it.

For example, look at the Level 2 PAs in the Project Management category, PP, PMC, and SAM, and RSKM in Level 3. It is relatively easy to write one or two organizational processes that *tell* people *what* they're supposed to do in these areas, but it's a different task to build things that *show* them *how* to do the project management work. In my work with organizations, the number of process implementation assets that usually need to be built far outnumber the number of processes. An incomplete list of candidate process implementation assets which could be built for the project management processes in our example includes:

- Project management process checklist
- Project initiation plan template
- Project scoping worksheet, guideline, or template
- ROM estimating worksheet
- Life cycle selection criteria and guidelines
- Project approval form
- Stakeholder plan template
- Project management plan template
- WBS template
- Effort and cost estimating worksheet, template, or form
- Risk planning and management template
- Project measurement subplan template
- Project measures collection worksheet
- Project configuration management subplan template
- Project quality assurance subplan
- Project status reporting template or form
- Project review pro forma
- Meeting agenda and minutes templates or forms
- Project action item template
- Project lessons learned pro forma
- Project close-out checklists

It's just not good enough to put processes and procedures into peoples' hands and expect them to go perform. Set up a process implementation asset team that is chartered with reviewing the processes and procedures as they are being developed, coming up with things that provide an interface between the processes and the people, and enabling people to perform the processes effectively and efficiently.

DO'S AND DON'TS

Now that you've read this chapter, you have hopefully learned some new ways to think about CMMI-based process improvement and maybe you have even challenged some of the perceptions or closely held beliefs you have had about the way you work with CMMI.

Here's the summary checklist of do's and don'ts to serve as a reminder of the major concepts presented in this chapter.

Do

- Before you apply CMMI-based process improvement toward achieving your organization's goals or addressing systems engineering problems, make sure you and all the decision makers truly understand the goals and comprehend the root causes of the problems.
- Recognize that CMMI may not be the best or only way to address organizational goals or resolve systems engineering problems.
- Recognize that when you use CMMI to implement process improvements, the product you will deliver is a system, specifically a process system, and that the project to deliver the process system will experience the same cost, schedule, and quality issues that are always faced by all systems delivery projects.
- When starting process improvement, look for existing processes, practices, and work products which show potential for being incrementally improved (nurtured) and grown into viable process assets as part of the CMMI program.
- Understand the different ways organizations and people learn new things. Realize that putting new processes and procedures into people's hands and expecting them to instantly perform those processes, which is online learning, is the most expensive and least effective form of learning.
- Think about the meta-components of CMMI for a more effective approach to designing, developing, and implementing the CMMI-based process system.

Don't

- Don't assume that CMMI is the panacea for all systems engineering problems or that it is the "silver bullet" for achieving the organization's business goals.
- If the organization applies CMMI-based process improvement to resolving software or systems engineering problems or achieving business goals, and it accomplishes neither, don't just keep doing more of the same.
- In delivering process improvement, don't play with the delivered quality of the process system. The only three areas in which you

really have room to move are cost, schedule, and the development process.

■ Don't assume that just because a practice, process, or work product exists prior to the introduction of CMMI in the organization that it isn't worth saving and using in the CMMI-based improvement effort.

■ Avoid following the traditional vertical or "silo" approach to process design, definition, and implementation.

WHAT DID YOU LEARN? WHAT WILL YOU DO?

Now take the post-chapter quiz in Figure 4.10, and think about what you've learned and how some of your views toward CMMI-based process improvement have changed. Think about what you will do with the information you've learned (and how it makes you feel).

1. **Which of the following statements are true (may be more than one answer):**
   a. CMMI-based process improvement will address all systems engineering problems.
   b. Applying the CMMI to some systems engineering problems may backfire and cause the problems to worsen or result in unintended consequences.
   c. Everything you need to know about CMMI implementation can be found in the CMMI itself or in books about the CMMI.
   d. As long as your process focus people create detailed processes, people will be able to use them.
   e. All of the above.

2. **True or False:** If you build work products such as templates or forms which include embedded procedural detail, you may not have to write a process description.

3. **The best way to implement CMMI-based process improvement is to:**
   a. Think outside the CMMI and look to other industries for ideas about introducing and institutionalizing improvements.
   b. Balance cost, quality, and schedule in managing the process improvement project.
   c. Use an integrated approach in process design, development, definition, and implementation.
   d. Make sure that CMMI-based process improvement is appropriate for the goals the organization is striving to achieve or the problems it is trying to resolve.
   e. Start improvement by looking at what already exists in the organization, weeding out the unproductive or wasteful practices and work products, and nurturing those that look like they could be fruitful.
   f. All of the above.

4. **The most important idea I learned from this chapter is:**
   _____
   _____

5. **I will apply this idea in my process improvement work by:**
   _____
   _____

Figure 4.10   Chapter 4: What Did You Learn? What Will You Do?

5

FIVE CRITICAL FACTORS IN
SUCCESSFUL PROCESS
DEFINITION

When you are deluded and full of doubt, even a thousand
books of scripture are not enough. When you have realized
understanding, even one word is too much.

— **Fen-Yang**

WHAT DO YOU THINK? WHAT DO YOU BELIEVE?

Take a minute and answer the questions in Figure 5.1. Then, once you've
finished reading this chapter, take the quiz in Figure 5.4 ("What Did You
Learn? What Will You Do") to find out how much this information has
helped you with your own CMMI-based process improvement. Remember,
these quizzes are like process improvement work, there is rarely a right
or wrong answer; there are only answers that best suit your organization's
business needs.

THE MODEL AND THE REALITY

It is interesting to observe organizations engage in CMMI-based process
improvement. People will spend tens of thousands of dollars creating and
maintaining or buying processes and procedures they believe are "compliant" with CMMI practices and, somehow, the act of creating tons of
procedures and paper becomes synonymous with process improvement.
It's almost as if for some organizations the measure that processes have

1. **True or False:** Writing processes and procedures which are based on the practices in the CMMI are an effective way to improve the organization.

2. **Which of the following statements are false (may be more than one):**
   a. There are many factors other than the CMMI to consider when developing new processes or procedures.
   b. People just need to know the CMMI to write processes and procedures.
   c. When designing processes and developing process standards, defining the content for various types of process assets is more important than format.
   d. Everyone prefers to have processes in graphical format such as flow diagrams.
   e. An organization should determine how people are going to use the processes and procedures as part of the design.

3. **True or False:** Processes will be more valuable and useful to people if they help them create the deliverables for which they are responsible.

4. **One of the most important skills sets to have on the team responsible for defining new processes is**
   _____.

5. **True or False:** A procedure must be a very long and detailed narrative document.

Figure 5.1   Chapter 5: What Do You Think? What Do You Believe?

been improved is the increasing page count of their process documents. The irony of this is that in all of CMMI, there is but one GP (GP 3.1), described in two-thirds of a page, that addresses defining and documenting the organization's processes.

There are two truths that organizations should know about process improvement. First, many management and engineering processes and practices can be improved without documentation. Second (since that first truth won't be widely accepted), the organization is bound to spend lots of resources documenting its processes and it should understand that there are effective and efficient ways to go about implementing GP 3.1. This chapter provides some process definition approaches and techniques which have worked and saved the organization both time and money.

WHY A CHAPTER ON PROCESS DEFINITION?

Process definition — the work of creating or revising organizational policies, processes, procedures, and other process assets — is just one small part of an overall process improvement project. So, why an entire chapter on the subject? That is a great question and one that made me reluctant to even dedicate a chapter to this topic.

Although process definition is just one phase of CMMI-based process improvement work, I have observed process definition work consume most or all of an organization's process improvement or CMMI budget. I've witnessed processes being defined in perpetuity and never being released into the user community for use. I've witnessed process definition work taking many months to complete, only to yield processes which were entirely unusable in the organization. Yes, process definition work should be just a small part of process improvement. This chapter reveals some techniques and lessons learned from practical implementation that you can use to ensure your process definition work is done with the appropriate priority and resources, which results in usable products that help people do their job more effectively and efficiently.

CRITICAL FACTOR 1: MAKE THE PROCESS WHAT PEOPLE DO

I worked with one organization which took the familiar path of setting up process improvement teams to define organizational processes. The teams were formed by and named after the PAs in CMMI. Each team was composed of 4 to 6 members and they all worked for about 3 to 5 months to develop organizational processes which were also named after the CMMI PAs. Each member of each team spent about 2 to 4 hours per week on his or her team's process definition tasks. So, if you make some conservative assumptions and do the math, it cost the organization some-where around 300 effort hours (probably about $15,000) for each of 6 defined processes. These estimates do not include building process imple-mentation assets such as forms, templates, or checklists, nor does it including piloting the defined processes in the organization; it is strictly process definition work.

The Supplier Agreement Management Team developed a 50-page process which was called — you guessed it — SAM. Upon completion, the team gave it to people in the organization who have the responsibility to purchase materials and negotiate and establish contracts with vendors, suppliers, and consultants. The purchasing and contract management people reviewed the process document and rejected it outright. Why? The documented SAM process did not come close to describing how purchasing and contract management worked in the company.

So what went wrong? The people on the SAM team were very bright, very hard working individuals. They had been trained in CMMI and they knew what SAM was all about. They knew what a SAM procedure needed to describe to satisfy a SCAMPI appraisal, so they certainly had written a "desired state" process.

Here's what went wrong. The SAM team didn't stop to understand and incorporate into the process description the work being performed by people in the company who held jobs related to Supplier Agreement Management. They created a process which described very well how SAM could be performed in the organization, but they didn't create a process that described what people really do.

Eventually, my company was called in to help out. We pulled together in one room most of the people in the company who were responsible for material purchasing or planning and executing subcontracts. I asked each of them to go to a whiteboard and draw or write (in bulleted tasks) the processes they used. In a 2-hour workshop on 1 afternoon, 5 people had defined the organization's SAM process. It took about another 60 hours to turn it into something usable but, the point is, we made the process what people do.

A process description — a document — is a virtual representation of a process; it is not the process. The process is what people do; it is the work they perform. I've seen appraisal teams come up somewhat empty handed in terms of finding documented organizational processes and conclude that "the organization has no processes." They were wrong. Every organization has processes; you just need to stop looking for paper long enough to see them.

The fastest way to define organizational processes is to go out and talk to people or observe them doing their job. Record what different people in the same role do; then look for the common tasks or activities and any common tools or work products they use. This will be the basis, the starting point, for the process definition — the paper. Of course people won't be doing exactly what's written in CMMI. They can't because CMMI is a model and it defines a desired state. The process descriptions and documents will need to be the bridge between what people are doing (the current state) and what the leadership wants them to do (the desired state).

(Also read Chapter 1 — News Flash! There Is a Level 1!)

CRITICAL FACTOR 2: PLAN PROCESS DEFINITION WORK

The second critical factor, and one frequently overlooked, is planning the process definition work. It is easy to forget about planning process definition; after all, it's just writing a bunch of procedures, right? Not exactly.

Remember, the process focus people such as SEPG or the senior management presumably believe in the principle that putting effort into planning will save time, money, and frustration later in the life cycle. (If the process people in your organization don't believe in this principle, then they need to stop pushing about five of the CMMI PAs on everyone else!)

The process definition phase of your organization's CMMI process improvement project deserves to have its own subplan, complete with a WBS, schedule, and resource allocations. This process definition subplan will, in turn, help the various process definition teams plan their respective work.

Not surprisingly, there are many similarities between the development of a software-intensive system and a process system. A good place to look in your organization for ideas for a plan for process definition is the project plans for systems development.

Defining and Estimating Process Definition Tasks

There is a set of tasks and activities which are almost always performed in process definition, yet few organizations actually plan these tasks and activities by allocating time and resources to performing them. Inevitably, the CMMI or process improvement project overruns cost and schedule because "of all those things we had to do which we didn't know about." So know about them now and build them into your plan.

The commonly performed yet infrequently planned process definition tasks include:

- Establishing the process definition teams (read "Integrated versus Vertical Approaches to Process Improvement" in Chapter 4 — Process Improvement Strategies that Work)
- Defining organization and process terminology (read "Critical Factor 3: Define the Process Language for Your Organization" in this chapter)
- Establishing format and content standards for process assets (read, "Critical Factor 4: Design First, then Build" in this chapter)
- Developing/designing use cases

- Establishing and defining process asset verification (aka peer review) criteria
- Defining process asset configuration management, version control, or release management standards and procedures

Some of the more important and less understood items in this process definition task list are expounded upon in the rest of this chapter.

Building the Process Definition Teams

Going back to our systems development analogy, would we expect the people doing requirements analysis to also do the implementation or coding and then also do the integration test? Would we expect software engineers to also perform all the project management functions, requirements management functions, and verification and validation functions? Probably not. So why do we expect a SEPG to have the skill, knowledge, and resources to perform all the tasks in the end-to-end development and delivery of a process system?

This really comes down to answering the questions:

- What is the difference between the SEPG (or whatever you call the people with process responsibility) and the CMMI process improvement project team?
- Do they have to be the same people?

Most organizations, even if they buy into the concept that process improvement can be "projectized," simply assume that SEPG is the CMMI or process improvement project team. But it doesn't have to be so. If you think about it in terms of predefined longevity, a process improvement project team needs to exist only for the duration of the project or release, say moving from one maturity level to the next. The SEPG, on the other hand, is usually an organizational function which has no predefined or preset duration for its existence. The process focus function is more of an ongoing interest in the organization, like the personnel department or the training department; they are the process department.

We can apply this model to CMMI-based process improvement. The project teams in your organization probably exist only for the duration of a project. During the life cycle of the project, the project manager acquires services and resources from the ongoing functions in the organization such as architecture and design, engineering, manufacturing, shipping, personnel, testing, etc. Similarly, the process improvement project team can be established to exist only for the duration of a defined process improvement project. This project team will also plan and then execute the project by acquiring and temporarily managing resources from ongoing functions such as SEPG, engineering, management, training, etc.

This is exactly what I did in CSC and it worked beautifully. I was the project manager for the project to move the organization from CMM Level 2 to CMM Level 3 and achieve some other defined business goals. The process improvement project team was a separate and distinct unit from SEPG and planned and used resources from SEPG, the quality assurance group, the configuration management group, and other ongoing concerns. This model can work very well in organizations that draw the distinction between a "program" and a "project," with that distinction being that projects have a finite start and end date and programs, which are often an ongoing collection of related projects, do not. In the CSC organization in which I worked, SEPG ran the process improvement program to which the process improvement project I managed was a contributor.

(For more information, read "Establishing the Process Improvement Project Team" in Chapter 3 — Managing the Process Improvement Project.)

CRITICAL FACTOR 3: DEFINE THE PROCESS LANGUAGE FOR YOUR ORGANIZATION

What we call things seems so trivial that we see no reason to waste time discussing language or asking much less answering questions like, "what do we mean when we use the word 'process'?" After all, everyone with whom I work has the same meaning and understanding of common words and phrases as I do, don't they?

But it's not trivial at all; in fact, it's big … very, very big! Not establishing common definitions and a common understanding of the difference between the words "policy," "process," and "procedure" can cost your organization tens of thousands of dollars. Not establishing common definitions and an understanding of the word "organization" as it is used in CMMI-based process improvement can cost millions of dollars. For example, let's say you and your SEPG work in a midsized IT shop in General Motors Corporation that is responsible for developing and maintaining a supply chain data warehouse. A vice president asks you to put together a project plan for CMMI-based process improvement for the "organization." You and the other process people build a plan for implementing CMMI-based systems engineering improvements in the supply chain data warehouse and present that plan to the VP. Then he says, "Oh no, I meant a process improvement plan for every software and systems engineering unit in GM." Your plan and estimates might need a little revision, eh?

The next two subsections give you some insight into the words and phrases that are the most important in terms of planning and performing process definition work. The most important thing to remember is this:

Your organization does not have to invent definitions for these words. There are plenty of references, such as the CMMI Glossary, that contain definitions which will give you a workable starting point. However, generic or textbook definitions will not be sufficient for your organization and its specific CMMI implementation and process improvement work. In the end, what really matters is that communication within your organization and with stakeholders is not misunderstood; that is the ultimate success criteria.

Defining Process Assets

Process definition is the phrase normally used to describe the activities involved in designing and developing policies, processes or procedures, and related process work products such as forms, templates, checklists, etc. In some shops, process definition can also include process change management and process deployment. And that's about all the consensus you'll find in this area. Figure 5.2 illustrates most of the terms related to process definition that people fail to define, which leads to waste and rework downstream.

Within the process improvement or CMM and CMMI industry, we use words such as policy, process, procedure, guide, handbook, template, checklist, and form frequently, but we often use these words without a consensus understanding of their meaning. You know what you're talking about when you use these terms. If you assume everyone else has the same definition, you'd be wrong. Test this out: Go out into your organization and ask three people (separately) the following questions:

■ What is a policy? How is it used and by whom?
■ What is a process? How is it used, by whom, and when?
■ What is a procedure? How is it used, by whom, and when?
■ What is the difference between these three types of documents?

If you get the exact same answers, write to me. It seems that in the area of process definition, some definition is very much needed. Take a look at Figure 5.3. It introduces some rationale for defining three of the major terms used in process definition and improvement.

In addition to policy, process, and procedure, you will also need to define and distinguish between "template" and "form." You may also have to define "guideline," "handbook," "criteria," and "desk instruction."

One process asset term that sometimes makes people get really testy is "plan." I have frequently come across documents which people referred to as a "plan." The so-called plans described how something was going to be performed, but it did not describe tasks, assumptions, dependencies,

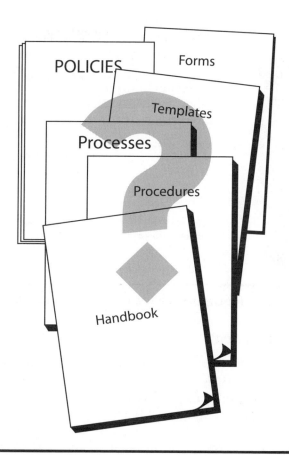

Figure 5.2 Process Asset Terms Which Often Do Not Get Defined

constraints, risks, effort and cost estimates, or schedule. My own experience and cognitive filters caused me to have a hard time accepting such documents as plans, but their owners were quite adamant that they were plans. Hence the problem when you don't have consensus definitions on such terms.

Defining the Language of Your Organization

In addition to defining terms that represent different process assets, the people involved in process definition work and the stakeholders thereof also need to define other words which are probably spoken and assumed, but undefined in the organization. Several terms are more critical to define than others. The following subsections identify these critical terms, why they're so important, and the questions people need to ask to arrive at consensus definitions.

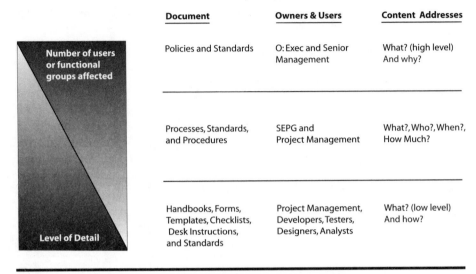

	Document	Owners & Users	Content Addresses
Number of users or functional groups affected	Policies and Standards	O: Exec and Senior Management	What? (high level) And why?
	Processes, Standards, and Procedures	SEPG and Project Management	What?, Who?, When?, How Much?
Level of Detail	Handbooks, Forms, Templates, Checklists, Desk Instructions, and Standards	Project Management, Developers, Testers, Designers, Analysts	What? (low level) And how?

Figure 5.3   Rationale for Defining Policy, Process, and Procedure

Defining "Organization"

This word is by far the most important word to define before any realistic planning for CMMI-based process improvement can be accomplished. Every aspect and parameter of a CMMI process improvement project — the scope, cost, schedule, quality, appraisal — everything depends on the definition of this one word. In planning and executing the CMMI process improvement project, everyone will need to know which organizational units will be involved, which people or roles will be affected, which programs and projects will be affected, which systems and products will be affected, and so forth. The definition of this word defines the boundaries of the CMMI process improvement effort. No one in the organization can afford to assume its meaning. Most of all, realize that for CMMI process improvement and an eventual appraisal, the "organization" may not be the same as the organization chart.

Defining "System" and "System Engineering"

Maybe your organization moved from SW-CMM to CMMI because the leadership realized the not-so-widely-acknowledged fact that there are really very few pure software organizations left in the world. Maybe, there never actually was a software industry because you don't have to think too hard to realize that software does not nor has it ever done anything by itself. Software has always been a part of larger systems — from wrist watches to space shuttles — that perform tasks; they do things. The more likely reason your organization is moving or has moved to CMMI is because everybody else is doing so. That's okay, but if your process improvement effort previously focused on software, the organization now needs to define what the word "system" means in the context of their business and process improvement work.

As with other words your organization needs to define, there are plenty of published references that can give you good definitions for a "system," but the only definition that counts is the one that works for your organization. By reading this book, you already have had your concept of "system" expanded to include process systems. Your place of work is a "system of systems." There are sociological (people) systems, there are technological systems, and there are process systems. All of these subsystems work together to make the organization a viable, functioning system.

However, for the purposes of CMMI-based process improvement, if you're using the systems engineering discipline (CMMI-SE/SW), your organization will need to define what "system" means in terms of improving system engineering processes. This task is anything but easy because, inevitably, the definitions people start talking about will challenge and threaten other people's long-held beliefs of the definition of a "system."

Defining "Project"

Watch out! This is another word that will get people riled up when you start asking what it means. I had a person yell at me, "I know a project when I see one and that's the only damn definition we need!" That attitude is perfectly workable so long as no one has to ever work with anyone else. The problem comes when two or more people have to work together on achieving a common task or goal — like what we do every single day in the business world.

The reason it is so critical to define the word "project" up front is because most of the CMMI practices in the Project Management and Engineering Process Area categories, and many of the practices in the Support Process Areas, are based on an organizational definition for "project." Your organization doesn't have to use the word project; it can call the same grouping of work by a different name. Many of Natural SPI's clients in the defense contracting world refer to "programs," which turns out to be their equivalent of what most of us call a project.

Earlier in Chapter 3, the text provided a starting-point definition for "project" which is worth repeating. Generically, a project can be defined as a collection of tasks or activities that:

- Have a start date and an end date
- Will deliver something that can be defined, measured, or tested
- Will be allocated defined resources
- Will deliver something to a customer or customers who are identified

That's easy enough, but eventually the process people will run into some difference of opinion with relevant stakeholders on the definition of a project. People will wonder how software or system maintenance fits into the definition of project. Let's say you have an organization that gets an annual budget to fix or repair system defects reported to it by users and customers. The organization resolves hundreds of defects a year and releases dozens of patches or software system releases which contain the solutions to the defects. You could tell these people that each individual reported defect constitutes a "requirement" and that a single solution to a single defect constitutes a "project," but you'll probably be inciting a riot by doing so. The people working maintenance or systems sustaining will be very vocal in pointing out the insanity of applying full project management discipline and practices to a single defect resolution and they would be correct.

The good news is that your organization gets to define "project," even if that definition differs from the industry's glossaries. If you have a maintenance or sustaining budget with which the shop will deliver some number of defect solutions to a defined customer, doesn't that constitute a set of defined tasks or activities with a start and end date, allocated resources, defined deliverables, and defined customers? Is that not a project?

That won't be the end of the dialog on this subject; it is not easy. Most likely the next series of conversations will be to answer the question, "what do we mean by 'development'?" because people make the reasonable assumption that the CMMI-SE/SW best fits software or systems development. This discussion lasted months in one of my client organizations. The question is very easy to answer if an organization is truly building a completely new system or product from scratch, but the answer gets elusive for questions such as:

- How much of an existing system can change before it is considered development?
- If fixing defects significantly changes the system, is that development?
- Is migrating a software system to a new platform considered development?
- Is rewriting a system in a new language development?

There are no easy answers to these and other questions. There is, however, only one right answer: the answer that best suits the needs of the people in your organization who will need to work with the answer in the future.

Defining "Stakeholder" and "Relevant Stakeholder"

One of the significant benefits of CMMI over predecessor maturity models is that it codifies useful guidelines for involving stakeholders and relevant stakeholders in processes and project decisions and work (GP 2.7). The downside, as usual, is that you and your organization get to do the hard work of figuring out what is meant by "stakeholder" and who are the stakeholders in different decisions and actions.

The importance of defining the stakeholders or relevant stakeholders for decisions and activities cannot be overstated. Here is just a short list of real situations that resulted in waste, rework, lost time, momentum, or emotional capital which were caused by not having defined these terms:

■ Processes and process assets thought to be complete had to be reworked because a stakeholder was not included in the review and approval.
■ Completed project deliverables were not accepted by a customer because he wasn't involved in defining the requirements or other project decisions.
■ A group of people ignored the new processes and work products because they didn't have a say in their design and development.
■ A project's scope, effort, cost, and schedule were woefully under-estimated because a relevant stakeholder was excluded from the project initiation decisions.

I'm quite certain you have many similar war stories of your own on this topic. Again, the CMMI Glossary and other references can provide you with a generic definition for stakeholder, which the process focus people need to take and modify to make it real and workable for your organization.

One lesson learned by organizations with which I've worked is that they needed to distinguish between the terms "stakeholder" and "relevant stakeholder." For "stakeholder," they accepted the CMMI Glossary's definition verbatim. However, they also don't use this term very often and only use it in a generic context when it is spoken or written. In most cases, particularly in the processes and implementation work products, the organization has made the effort to define (by name, by function, or by role) the specific relevant stakeholders for a specific decision or action. For example, a decision to accept and baseline project requirements will have a group of relevant stakeholders identified for that particular decision. Within the same project, different relevant stakeholders may be defined for receiving project status reports. Natural SPI has provided some very convenient tools for documenting relevant stakeholders such as roles, responsibilities, and required skills matrix and a review–act–change–initiate (RACI) matrix. The mechanics of documenting relevant stakeholders is easy; building consensus on those decisions is the hard work.

The last important point to be made about defining stakeholders and relevant stakeholders is that stakeholders should never be defined and planned without those people agreeing to their involvement as a stakeholder. Some people may not want or think they should have participation in a certain decision or action and, if so, others should try to encourage and persuade the needed involvement.

Documenting the Organization's Language

Once the organization has defined its language and has obtained general acceptance of those definitions, this body of knowledge needs to be documented and disseminated to everyone using the language. This document will essentially be a translation table that provides the translation between "outsider" language, such as standard industry terminology, and the "insider" language. Table 5.1 shows an example of a partial language translation table.

Critical Factor 4: Design First, Then Build

As briefly addressed in Chapter 3 — Managing the Process Improvement Project, process systems are sometimes built without an architecture or design. Not surprisingly, such process systems are often later rebuilt because the builders, usually SEPG, skipped design decisions which would have yielded a system closer to correct the first time.

In our work, we have found there are at least four (probably more) significant components or subsystems of a process system's design. These design components need to be thought through, discussed, negotiated, and documented before people go off and write processes or procedures. The four main process system components are:

- *Content:* Which elements of information does each process asset type or class contain?
- *Format:* What do the various process assets look like?

Table 5.1   Sample Organization Translation Table

Outsider Term or Phrase	Our Organization's Term or Phrase	Our Definition
Project	**Task**	A task is a unit of work or a logical grouping of units of work requiring a total estimated effort greater than 120 hours or an estimated duration greater than 3 weeks. A task must also have defined or definable requirements.
System	**Products**	Products are deliverable physical items or services produced for the customer.
Senior Management	**Program Managers**	For the purposes of systems development and delivery and process improvement, our program managers constitute "senior management." This includes grades E10 through E13.
Development	**Engineering**	The term "engineering" in our organization is used to describe all work and activities related to the development, repair, and maintenance of hardware, software, firmware, and products composed of any combination of hardware, software, or firmware.
Quality Assurance	**Quality Control**	In our organization, quality control (QC) is the same function as that described for PPQA in CMMI.
Configuration Management	**Change Management**	Change management is the term we use to describe all forms of establishing and controlling configurations of products and product components. Change management also includes the management of document revisions or versions.

- *Quality:* What testable criteria constitute "usable" or "good enough?"
- *Environment:* Who will use which process assets and how will they be used?

These design components and considerations are addressed in some detail in the following sections.

Process Asset Content

Let's go back to our discussion about defining what we mean when we use words such as process, procedure, policy, handbook, form, etc. In the section on Critical Factor 3 in this chapter, Figure 5.3 provides you with some rationale for distinguishing between three common process assets — policy, process, and procedure — and how each of these three assets are related. However, when it comes down to creating such assets, the developers of the process system are going to need more detailed guidance than that.

You could repeat the two most common mistakes made by organizations in designing process asset content:

1. Skip the discussions altogether, build the process assets, and then rebuild them when everyone realizes they should have a structured design.
2. Put a bunch of expensive people in a room and let them argue about it for days on end.

Or you can skip both expensive lessons, build something similar to the matrix shown in Table 5.2, and give people a straw-man content standard to jump-start this design component.

Table 5.2 shows an example of just one way to design the content for various types of process assets. Each column (following the first column) identifies a type or class of process asset such as process, form, template, policy. Each row identifies a content element, which often represents a section or subsection of information in a document. Where a column (process asset) and a row (process element) intersect, you find information that provides the content standard. Of course, providing this type of matrix alone is not sufficient. Each of the process elements (in the first column) must be defined so that people won't have to guess at their meaning as they create the process assets.

If the people designing the process system build something like Table 5.2 first, then people won't waste their time trying to figure out which types of information goes in each of the types of process assets, nor will the organization lose time and money rebuilding process assets which all come out inconsistent because there was no content design.

Format and Style

Designing the content of the organization's process assets is the most critical aspect of designing the process system, but designing the look and feel or presentation of those assets is also important. Essentially, you want customers and users of the process system to be able to look at

Table 5.2 Example of Defining Process Asset Content

Process Assets Content Definition Standard

Content Element (Section)	Policy	Process	Procedure	Plan	Template	Guide	Charter	Process Asset n
Governing Policy or Reference		✓			✓		✓	
Requirements				✓				
Audience/Users Description	✓	✓	✓	✓	✓	✓	✓	
Applicability/Scope	✓	✓	✓	✓	✓	✓	✓	
Required Skills and Knowledge		Suggested	Suggested					
Roles and Responsibilities				✓			✓	
Resources		Suggested	Suggested			Suggested		
Definitions	Suggested	✓	✓	Optional				
Entry Criteria and Artifacts (Inputs)					Suggested		✓	
Tasks/Steps/Activities (Procedure)	✓	✓	✓		✓	✓	✓	
Exit Criteria and Guidelines (Outputs)					Suggested			
Assumptions							Suggested	
Estimates		Suggested	Suggested	✓				
Schedule/Milestones				✓				
Dependencies/Constraints	Suggested	Suggested		✓				

(continued)

Table 5.2 Example of Defining Process Asset Content (Continued)

Content Element (Section)	Process Assets Content Definition Standard							
	Policy	Process	Procedure	Plan	Template	Guide	Charter	Process Asset n
Decision Criteria and Methodologies						✓	Suggested	
Process Relationships and References	Suggested	Suggested	Suggested		✓	✓	✓	
Exceptions/Waivers	✓	✓	✓				✓	
Illustrations/Flow Diagrams		Optional	Optional	Optional	Optional	Optional		
Subplans: Training, QA, CM, Safety				✓				
Review and Effectivity	✓	✓	✓				✓	
Revision History	✓	✓	✓	✓	✓	✓	✓	
Metadata	✓	✓	✓	✓	✓	✓	✓	

one component (i.e., one process asset) and intuitively recognize it as belonging to the larger process system. Accomplishing this goal requires that the process definition plan allocate some resources to designing process asset format.

There are several techniques for expeditiously designing and implementing process asset format which can save the organization time and money. Some combination of these techniques can be tailored to suit your organization's process definition needs:

- Assuming most of the process assets will initially be in the form of physical documents, recruit a publications designer to work with the rest of the process definition teams to create a consistent look and feel for the assets. Chances are pretty good that there is someone in your organization with this skill and experience, but you won't know unless you ask.
- Assuming most or many of the process assets will eventually be accessible electronically (i.e., through a Web interface), recruit someone from within your organization who has experience with designing online documentation systems. Have this person work with both the people doing process definition and the physical document designer (in the preceding point) to design the process system.
- For every type of process asset, design and build a template or a model of that process asset before anyone starts developing instantiations of the asset. For example, design and build a process template to give to process definition teams to use for building their respective processes. If MS Word is the tool being used to document the process assets, make sure the templates contain predefined styles for elements such as section headings, body text, figure and table titles, notes and cautions, and headers and footers. Predefined styles will save many, many hours of work for people who are very knowledgeable in the processes, but who are not experts in the use of document production or publication tools.
- Have each process definition team build only one instantiation of a process asset first: a prototype, if you will. Schedule and conduct some form of verification such as peer reviews for the purpose of verifying each process definition team's first output conforms to the approved design standards. This way, you catch the significant deviations and defects in the process assets before those defects are repeated and perpetuated in all the subsequent assets that will be built.

Process Asset Quality

Another area that usually becomes problematic to people doing process definition work is quality. The reason it becomes a problem is because the relevant stakeholders fail to define "quality" for the process assets they intend to build. It is ironic how quickly system engineering people can readily quote you the quality standards and thresholds for their system products in terms of defects, defect density, or defect severity. Yet, put them on a SEPG and ask them to define quality standards for process documentation, and they're suddenly and mysteriously mute.

What we've witnessed happen time and time again in organizations is that a bunch of people will build a bunch of process assets. If they're thinking, they will peer review or inspect those process assets before piloting them in the organization (never ask your customer to test for you unless you offer them incentives for doing so). This is where it goes south. In the first round of peer reviews, they'll find and remove from the process assets about 80 percent of the defects (deviations from standards and design). For some, this won't be good enough. So then, they'll conduct several more rounds of peer reviews or inspections, spending three to five times the amount of effort and money spent on the first review, to find and remove the remaining 20 percent of the defects. People will spend a dollar to find and correct a major process content defect, such as a workflow sequencing problem, and then spend five more dollars to find three typographical errors in a 25-page procedure.

Why do people do this? Well, some of it is the nature of people. There are those among us who continually strive for perfection even when perfection is not defined and therefore, by definition, not achievable. There are others among us who simply cannot leave a document untouched; they must make some change to it no matter how trivial because then they will feel they've contributed. Yet other people who review a process asset are not really qualified to make a substantive or contextual change, so they make unsubstantial changes because they don't know what else to do. Another reason is that we sometimes recruit highly technical people for our SEPGs and process definition teams. Engineers and software developers come from a world in which one singular mistyped character in one line of a software program can cause the entire system to fail. Such people have difficulty changing their frame of reference to one in which it's almost impossible for a singular typographical error in a process document to cause the process to fail.

However, experience and observations tell us that the number one reason people spend too much money on their quest for quality in process assets is because they forgot or didn't think to define quality up front, before the process assets were built. In other words, they didn't define "good enough" — good enough to pilot, good enough to implement, good enough to train, and so forth.

Yet defining quality for process assets is not so difficult. It can be predefined in either qualitative or quantitative terms or in both. A quantitative definition for quality in a process system is similar to the quality parameters a project and stakeholders establish for other types of systems. Classes and severity of process defects can be defined (at least by example if not in abstraction). Then numbers of defects or defect density percentages can be established for "passing" verification or validation or for releasing the process system products into the field for piloting and use.

Quality for the process system can also be qualitatively predefined in terms of risk. The question, "is it good enough?" can be answered by the answer to the question, "what problems will the unresolved defects cause if we release the process asset 'as is'?" People involved in process definition and design can even roughly estimate the cost of releasing process assets containing nonsevere defects and determine an acceptable cost, below which it doesn't make financial sense to perpetually pursue low-risk defects. Or, the decision makers in this phase can simply use some broad-brush questions to determine the risk of defects in the process assets such as:

■ Are the defects so severe that they will cause work stoppage?
■ Will the defects cause the process to be performed incorrectly?
■ Will the defects, although individually minor, when viewed in the aggregate result in an overall unfavorable impression of the CMMI process improvement project in the community?

The people designing the risk thresholds can decide that if the answers to all of the above questions are "no," then the process assets will be released to pilot or to implementation and the defects will be removed later as time permits.

When implementing improvements in the organization, it's important that the people responsible for those improvements always seek an appropriate balance between making their products good and making their products available to the user community. If the process improvement project team or the process focus people hold onto the improved process assets too long trying to make them perfect, their customers — the users of the processes — won't wait. The users will keep doing their work the way they always have, and they won't be very forgiving that the process improvement project seemingly doesn't have to hold to a schedule while they do. On the other hand, if the process improvement project prematurely releases unusable junk into the user community, project improvement will probably never regain the trust and faith of the users. So, it is critical that people involved in planning and designing the process system establish the criteria for "good enough" early in the process improvement life cycle.

Process Usage and Customer Environment

The fourth critical design component of the process system concerns defining the environment in which the process system will be used and designing how people will use the system's components and products. The key questions which need to be answered to inform these design considerations are:

- How will people use the process assets? Will the process assets be used in physical (paper) form or will they be used electronically?
- How will people want to move from one process asset to another when the assets are linked?
- How much detail is needed for different classes or types of users?
- How will people navigate through the process system so that they can quickly get to the particular asset or task they need to perform now?

The process improvement project team and the process focus people, as smart as they may be, probably shouldn't try to answer these design questions in a vacuum. Pull together some small focus groups of potential users of the process system and get their answers to these questions. If possible, show them some existing or demonstration process systems and get the users comments on what they like or don't like.

The one aspect of this area of process design that I have found to be overlooked and underestimated most often is the consideration of how much detail is needed for different classes or types of users. People involved in process definition don't ignore this concern, but they do struggle with it. In the end, thinking they have to decide between too much process detail and too little, they almost inevitably settle on providing too much. The resolution to this dilemma does not have to be too much or too little. A process cannot be all things to all people, but it can be many things to many different levels of users.

Let's look at a project planning process as an example. In any random population of a software or system engineering organization, there will be almost infinite different levels of knowledge and experience in project planning tools and techniques, and there will certainly be many different needs for the project planning process and its resulting work products. To illustrate how a process can be designed to satisfy multiple levels of experience and knowledge and different levels of need, we'll define the superset of potential users and stakeholders of the project planning process into five subsets:

1. Novice project managers who have very little knowledge of or experience in project planning processes, techniques, or tools.
2. Intermediate project managers who have pretty good knowledge of and experience with how the organization does project planning.
3. Expert project managers who have extensive knowledge of and many years experience in project planning processes, tools, and techniques (i.e., a PMI-certified PMP).
4. Project team members such as analysts, developers, engineers, and testers who contribute to project planning at the direction of the project manager, but who do not directly use the project planning process.
5. Stakeholders such as program managers, senior managers, and customers who are usually recipients of the outcomes and outputs of the project planning process being performed.

Table 5.3 lists each of the classes of project planning users, describes their respective needs or levels of interaction with the process, and then provides some process design concepts for providing a process solution for each class of user.

These same design concepts can be used to design other organizational processes and process assets, and it doesn't necessarily mean the development of multiple documents. For example, for every task or activity statement in a process description that describe the "what" to do, you can embed a link or a macro labeled "show me how" that takes the user to the next lower level of procedural detail, if needed.

The point being made here is that one level of detail does not fit all; it never has and it never will. The people in charge of designing and developing process assets don't have to choose between alienating a particular class of user to appease another.

CRITICAL FACTOR 5: FOCUS ON PROCESS IMPLEMENTATION ASSETS

This critical factor was briefly addressed in the section, "Work Product-Based Approach to Process Improvement," in Chapter 4 — Process Improvement Strategies that Work and here we expand on this concept.

The topic is really a discussion of philosophies; it's a discussion of what process-focused people tend to do versus what delivery-oriented people do. Hard-core process people, such as the people you find on SEPGs and at the SEPG Conference (including this book's author), tend

Table 5.3 How to Design Process Assets for Different Classes of Users

User Classes for Project Planning Process Assets	Information Needs and Process Interaction Level	Process Asset Design Concepts
Novice project manager	Needs lots of procedural "how to" detail and self-instructing templates, checklists, and forms	• Planning process; "what" needs to be performed • Detailed "how to" planning procedure • Self-instructing project plan template • Self-instructing templates, forms, and worksheets for estimating, risk analysis, and resource planning
Intermediate project manager	Needs mostly process "what to do," does not need procedural detail for project planning	• Planning process; "what" needs to be performed • Self-instructing project plan template • Self-instructing templates, forms, and worksheets for estimating, risk analysis, and resource planning
Expert project manager	Needs quick, easy tools for creating and delivering project planning work products; does not need process-level or procedural detail	• Checklist of major project planning tasks and deliverables • Self-instructing project plan template • Self-instructing templates, forms, and worksheets for estimating, risk analysis, and resource planning
Project team members	Need awareness-level knowledge of which aspects of project planning to which they contribute; need self-instructing templates or worksheets for project estimating	• Checklist identifying project planning tasks in which they are involved • Self-instructing templates, forms, and worksheets for project planning tasks to which they contribute
Project planning stakeholders	Need awareness-level information on work products and other planning outputs they should expect to receive; also need summary of their participation as a stakeholder in project planning	• Checklist of project planning outputs they should expect to receive and instructions on what to do with those items • Project roles and responsibility matrix or stakeholder participation matrix

to care a lot about how things are done, but sometimes forget that performing a process is supposed to yield something, like a product. It is the means that's important, not so much the end result. Delivery-oriented people, which include people in project management jobs, sales and marketing, and engineers are producers. They want to create and deliver something and how they get that done is not as important as the results. These people are sometimes happy creating things no one needs or wants; they just need to create and produce. These are broad, sweeping generalizations which you may think are unfair, but pop your head up and look around and I think you'll make similar observations.

The importance of this observation of philosophical approach is that it is one of the reasons process people (like you?) in the organization often fail in their mission. It is also one of the root causes of the frequently observed contention between people in process roles and the rest of the organization. It is critical that you think about these differences in the way people see the world if you want your organization's CMMI process improvement effort to succeed.

And while you're at it, think about what business you're in — assuming you are involved in CMMI or process work in your organization — and answer this question: If the business of process is defining the way people work and the business of process improvement is helping people improve the way they work (i.e., helping them be more efficient, effective, and perhaps happier), which business are you in?

Organizations can and often do build processes based on CMM or CMMI. In doing so, they define the way work gets done, but they do nothing to improve the way work gets done. Yet other organizations improve the way people work even though they've never even heard of CMMI. Again, in which business are you? In which business do you want to be? Choose and choose wisely.

At this point, some of you and especially those of you who hate philosophical discussions, are asking, "what does any of this have to do with the critical factor of focusing on process implementation assets?" Keeping to one of the principles in this book, I'll show you, not just tell you.

Give the Deliverers Deliverables

If you are in the business of process improvement, then presumably you know your customers and what they want. And if you know this, then you know they want to get their work done as quick as possible, with as much quality as possible, collect their pay, and go home. What are you doing to satisfy those needs?

If you give an engineer a procedure for analyzing requirements (RD), all you've done is define the way you think CMMI wants her to work. If you give her a checklist or Excel spreadsheet in which she can plug in a few answers and the tool spits out a determination on whether the requirement is acceptable or not, *now* you've helped her do her job as quickly as possible, with as much quality as possible, collect her pay, and go home.

If you give a project manager a long, detailed procedure for developing project effort and cost estimates, all you've done is define his work in terms of CMMI. But give him an MS Word template with macros which enable him to answer some questions about the project and then the project plan practically writes itself, *now* you've improved the way he works!

If you give the quality assurance specialists a procedure that waxes philosophical about the merits of CMMI-compliant process and product quality assurance, all you've done is either irritated them or put them to sleep. But put one-page quality assurance checklists in their hands and *now* you've helped them do their job more efficiently and accurately. And, you've turned them into strong supporters of the CMMI process improvement initiative.

Let the process people build their processes and procedures if they must, but give the deliverers their deliverables. Build process implementation assets, which are deliverables that, by their content, are inherently consistent with the organizational processes so that when people do their work and produce things, they are automatically compliant with the processes without even thinking about them.

Show, Don't Tell

In the same spirit of giving the deliverers their deliverables, there is another critical concept to incorporate into process definition work: the concept of showing people what to do and how to do it, not telling them. This doesn't mean simply replacing process words with pictures; it means demonstrating for people what they're supposed to do, and how they're supposed to do the task. So let's practice this principle together by illustrating the concept.

One of my favorite PAs in CMMI is Measurement and Analysis (MA) because it's such a gorgeous PA, yet it is so often poorly implemented by organizations. The organization's process definition team can easily write a procedure which more or less mimics the MA practices. The result will be a nice looking process definition (GP 3.1; a process) to show to an appraisal team, but which does nothing in terms of helping people perform in this area.

Conversely, you could take the approach which has been enormously successful for Natural SPI's clients. We conduct a two-hour workshop based on Goal-Question-Metric,[19] which we have trademarked as "GQM Lite." This workshop walks people step-by-step through defining and prioritizing the organization's business goals (MA SP 1.1). The workshop explains how to translate goals into measurable concepts or indicators, which in turn are supported by derived and base measures. We then give them a blank form (implementation asset) to complete for each goal. We also give them a sample form that has been completed to use as a model (implementation asset) so that they can easily complete their forms. The Natural SPI GQM Lite form, which usually takes people less than one hour to complete when they have an example to reference, asks the following questions in natural language:

- Given the goal, what questions would the organization have to ask and answer to know if the goal is being achieved? (MA SP 1.1)
- Given the questions, what are the measurable concepts or indicators which would quantitatively or qualitatively provide the answers? (MA SP 1.1)
- Given the measurable concepts, what are the derived measures that would provide the indicators? (MA SP 1.2)
- Given the derived measures, what base measures need to be collected? (MA SP 1.2)
- Who will collect the measures, when, and how? (MA SP 1.2)
- What are the analysis techniques or methods to be used to analyze the measures? (MA SP 1.4)
- To whom will the measures, indicators, and information be reported and for what purposes or uses? (MA SP 2.4)

Immediately after these forms are completed, the organization can begin collecting the defined measures. Again, we don't tell them how, we show them how. We build a fairly simple Excel spreadsheet (implementation asset) that incorporates the responses to the GQM Lite form as outlined above. One spreadsheet in the workbook is blank to be used for collecting actual measurement data and a second spreadsheet in the workbook shows examples of rows, columns, and cells filled in with data so that users can *see* what is being asked for.

Within days — not weeks, months, or years — the organization is implementing processes consistent with most of the practices in MA. And here's the kicker: they're implementing the process without a process description document! They can write that later; they've got real work to do and the process focus people have helped them do it more efficiently and effectively. They can get their work done, deliver products and information, collect their pay, and go home. Ah, success!

DO'S AND DON'TS

Now that you've read this chapter, you have hopefully learned some new ways to think about process definition. Maybe you have even challenged some of the perceptions or closely held beliefs you have had about how processes are designed, developed, and implemented.

Here's the summary checklist of do's and don'ts to serve as a reminder of the major concepts presented in this chapter.

Do

- Make the process what people do.
- Plan process definition activity as you would any other aspect of the process improvement project. Realize that process definition is a significant and critical phase of CMMI-based process improvement. It is an area in which many organizations overrun process improvement budget and schedule.
- Establish the process improvement team such that its membership is somewhat fluid. Plan to bring different skills and knowledge onto the team at different times to have the skill and knowledge needed for a particular phase, task, or activity.
- Understand the skills and knowledge needed to do process definition work and make sure those people are involved in this work.
- Specifically and overtly define the words and phrases that will be used to define, design, and implement the organizational processes. As much as possible, build consensus understanding, if not concurrence, on these definitions.
- Plan and allocate time, money, and effort to designing the process assets before developing them.
- At a minimum, design the content, format, quality, intended environment, and use for the process assets.
- Make sure the process definition activities and work focuses on designing and developing process implementation assets such as job aids, templates, forms, and checklists. These things, more so than processes or procedures, are what helps people improve the way they work.

Don't

- Don't make the written document the process.
- Don't let people jump into creating processes, procedures, and process assets without first estimating and planning this phase of the process improvement project.
- Don't assume that the right people to manage the organization's overall process improvement program or CMMI program (i.e., SEPG) are necessarily the right set of skills and knowledge needed for process definition or other life cycle phases of the process improvement project.
- Never assume that people have the same meaning for commonly used words and phrases. Misunderstanding, if left to be perpetuated in the organization's processes, will result in significant waste and rework later on.
- Don't let the process focus and definition people deliver only processes and procedures. Not only will this not improve the organization's maturity or process capability, it will alienate the user community because it does not help them do their work more effectively and efficiently.

WHAT DID YOU LEARN? WHAT WILL YOU DO?

Now take the post-chapter quiz in Figure 5.4 and think about what you've learned and how some of your views toward process definition have changed. Think about what you will do with the information you've learned (and how it makes you feel).

1. **Which of the following statements are true:**
   a. A process is a document that describes the way people perform certain work in an organization.
   b. A process is the way people perform certain work in an organization.

2. **True or False:** The best place to start when defining organizational processes and procedures is to read the CMMI and see what it says to do.

3. **Which of the following skills and knowledge is not required for process definition work:**
   a. CMMI knowledge
   b. Information and knowledge capture skills
   c. Interviewing techniques
   d. Sales and marketing skills or experience
   e. Database knowledge or experience
   f. Design concepts
   g. Publications and document management skills
   h. Presentation and public speaking skills
   i. Extensive writing skills or experience
   j. All of the above skills or knowledge are required for process definition work.

4. **The most important idea I learned from this chapter is:**
   _____
   _____

5. **I will apply this idea in my process improvement work by:**
   _____
   _____

Figure 5.4   Chapter 5: What Did You Learn? What Will You Do?

6

ACQUIRING PROCESS
EXPERTISE AND TOOLS

There is nothing so dangerous as a good pitcher with no real
talent.

— **Roberto Goizueta**[39]

WHAT DO YOU THINK? WHAT DO YOU BELIEVE?

Take a minute and think about your experiences with (or maybe just
conference brushes with) me, your process improvement consultant. What
do you think is true or not true about hiring process or CMMI experts?
What is true or not true about buying tools to help your organization with
CMM or CMMI implementation? Take a minute and answer the questions
in the quiz in Figure 6.1. After reading the chapter, answer the questions
in Figure 6.6, "What Did You Learn? What Will You Do?"

THE MODEL AND THE REALITY

In CMMI, the process areas SAM and ISM provide guidelines for establishing and implementing processes for managing the relationships with
suppliers of products to the project or organization. It is easy enough to
understand how the practices in these two process areas apply to software
and systems projects, yet their applicability to other types of acquisitions
is not so obvious. With a little imagination and courage, organizations can
also use the principles and practices in these two process areas to improve
their management of the relationships with the suppliers of services. In
this regard, one of the fundamental philosophies underlying this book is

1. **Which of the following are good reasons to hire CMMI expertise or to buy a process implementation or improvement tool:**
   a. We're in a blind panic and we need help right now.
   b. My boss told me to go out and subcontract a consultant.
   c. It is more cost effective to hire outside expertise for specific types of process improvement work than it is to develop that expertise internally.

2. **Which of the following are good reasons for selecting a particular process consultant or process tool or system:**
   a. My friends and colleagues told me they were really good.
   b. They had a big booth at the SEPG Conference.
   c. They gave us verifiable performance data from their work with other clients.
   d. What they offer fits our requirements.
   e. Their approach aligns with our philosophy and culture.

3. **True or False:** Consulting is very different from CMMI-based systems engineering and delivery, so the CMMI doesn't apply to consulting work.

4. **Which of the following statements about your organization's relationship with a CMMI consultant or process tool vendor are true:**
   a. We need to accept the fact that there will always be CMMI knowledge or experience which we will need to hire from outside the organization.
   b. One of the ways a consultant or vendor adds value to the organization's process efforts is by helping us learn the things they know.
   c. There is never ample justification for hiring a consultant; we should have all the knowledge and experience in-house to do the job.

5. **True or False:** Hiring a CMMI expert consultant or buying a process tool has no relationship with the Supplier Agreement Management (SAM) Process Area.

Figure 6.1   Chapter 6: What Do You Think? What Do You Believe?

revealed: Process capability and organizational maturity is good not only for software and systems engineering, it is also good for every aspect of the business and every endeavor in which the business engages.

This chapter explains how to use the principles of acquisition management (SAM and ISM) to improve the acquisition of process expertise and tools.

PROCESS IMPROVEMENT — MAKE VERSUS BUY

The growth in the model-based process improvement business has given rise to an increase in the number of individuals and companies providing "expert" consulting in CMM or CMMI implementation process improvement. As of August 2003, there were over 2000 process improvement consulting firms or tool vendors which could easily be found on the Web. There are at least 6 after-market CMMI books available, the authors of which (including this one) would be happy to sell you advice, I'm sure. Organizations pursuing CMMI maturity levels spend anywhere from $5,000 to $10,000 per employee per year on CMMI-based process improvement work.[41] Major IT outsourcing firms such as SAIC and CSC and major "integrated solutions providers" such as Accenture market the rapid achievement of CMM and CMMI maturity levels as a significant market differentiator (can't figure that one out since they're all doing it) to secure multimillion dollar contracts.

Choosing the right consultant or solutions provider is difficult and risky and there just isn't much factual information to help you. With hourly rates for process consulting as high as $250 per hour and the cost of recovering from bad advice running that cost up many magnitudes, you can't afford to make too many wrong decisions about acquiring outside expertise.

Although there is no "Consulting CMM" to use as a guide for your acquisition decisions, there are some factors that will help you reduce your risks with process consultants and vendors. As with most acquisitions, hiring a process consultant or vendor should be based on verifiable and testable selection criteria. So, if your only criteria is to hire the friend of your golfing buddy, then stop here and good luck to you; otherwise, read on![40]

PROCESS IMPROVEMENT CANNOT BE OUTSOURCED

Whoa! There's an assertion that's going to make a lot of people unhappy. That's too bad; it's true and here's why:

- It is the organizational unit that gets appraised, not the temporary experts or "big-bat pinch hitters." By the definition of "institutionalization," the organizational unit must own the change, must own the CMMI-based processes. You can no more outsource process improvement and take the maturity level claim than you can hire expert interviewees from outside the organization to sit in on the appraisal interviews, although I suspect that has also been tried. Smart appraisal teams can figure out where the processes really live.
- Having tools that create all the necessary artifacts and process assets doesn't prove people are actually doing the work.

What an organization can do is prudently subcontract expertise or purchase tools which are cost-effective in building and implementing processes, but which also ensure the knowledge and technology is transferred into the organization to the extent that it becomes indigenous.

SOME REALLY BAD REASONS ORGANIZATIONS ACQUIRE PROCESS EXPERTISE OR TOOLS

Consultancy is very big business. Natural SPI's clients have spent anywhere from 5 to 20 percent of their entire budget allocated to CMMI-based process improvement and I'm sure they got every penny's worth! Chances are pretty good that someone in your organization will also have the idea to buy some outside expertise or tool to help with the CMMI effort. There are some really good and really bad reasons for outsourcing some aspects of the organization's process improvement work; let's start with the bad reasons.

Bad Process Acquisition Reason 1: They Had a Cool Booth at the SEPG℠ Conference

Well, okay, so they had a cool booth and they gave away some cool prizes or they engineered a very impressive product demo. Their full-color brochure is slick, their Web site incorporates all the latest Web technology, and they wear nice suits. Maybe everyone recognizes the company name or logo. And all of that has what to do with CMMI-based process improvement?

I remember attending a presentation at SEPG 2003 delivered by a vice president from one of the biggest names in management consulting. They were one of the conference anchors and had provided a keynote speaker. In the presentation, she boasted about how people in her organization worked nights and weekends to achieve their CMMI Level

3 target. Now wait a minute, let me get this straight, an accolade about process improvement being achieved through the heroics of individuals? Does anyone other than me see something wrong there?

If you choose a consulting firm or a tool vendor, just make sure your decision is based on actual performance data or experiential information, not solely on name or reputation.

Bad Process Acquisition Reason 2: They Guarantee Your Organization Will Be Certified at CMM/CMMI Level in X Months

There are lots of problems here. Let's start with the simple, provable fact that there's no such thing as maturity level certification. (Read Chapter 8 — Process Improvement Myths and Methodologies.) If someone is promising that, how well do they really know this business? Let's go a little deeper. Even if the consulting services or tool could ensure your organization achieves a CMM or CMMI maturity level within a certain time frame, is that the same as a guarantee that your organization will actually improve its overall systems engineering and delivery performance? You don't have to choose between getting a maturity level and achieving measurable improvement, but you do need to realize that the two are not always the same thing. Read the fine print or the lack thereof.

Bad Process Acquisition Reason 3: They're Great Golfing Partners

Or bowling buddies, or drinking pals, or whatever. There is a certain segment of the population for whom relationships are everything and facts such as actual performance don't even figure into decisions. I once had a boss who would come to my cube and, with great exuberance, announce to me a new member to the group. The conversations went something like this:

> BOSS: Hey, Michael, I just hired so-and-so to help out with your process improvement project!
> ME: That's great! What does he do?
> BOSS: Oh, he's a really great guy; you'll like him.
> ME: Okay, good. What kind of experience does he have?
> BOSS: He's really good … gets along with everybody.
> ME: What are his skills?
> BOSS: Okay, so teach him what you need him to do; he'll be great.
> ME: Sigh!

Deals, sometimes big deals such as consulting agreements or the purchase of a several hundred thousand dollar tool, will be made almost solely on the basis of a personal relationship. Sometimes, through pure dumb luck, such deals work out. Most of the time they don't because they're not based on the factual needs of the organization or on the vendor's historically demonstrated ability to deliver.

Bad Process Acquisition Reason 4: They're Cheap

Cost is always a factor in acquisition or procurement decisions; it should be. However, if cost is the only vendor or consultant selection criteria, then by the definition of goal and requirements traceability, the organization does not have any goals for quality, schedule, or approach.

Bad Process Acquisition Reason 5: They Used to Work at _____

You fill in the blank: SEI, IBM, Hill Air Force Base, Loral, the Software Productivity Consortium, Space Shuttle software, etc. Yes, there are organizations in the world that have been centers for brilliance, excellence, and innovation in process improvement and CMM and CMMI implementation. But as the stories grow older, they grow bolder, and now the legends are larger than the current-day reality.

The lesson is that good consultants can come from anywhere. They don't have to come from a place you recognize from the urban myths of process improvement to have good ideas and good experience which can help your organization. Just because a consultant or a tool vendor does come from a place of process legend doesn't mean that your organization will get what it needs and wants. Make sure the people in your organization who are responsible for making decisions about process consulting or tools can distinguish between reputation and demonstrable performance.

WHY BUY INSTEAD OF MAKE

There are really only two factors that provide ample justification for purchasing process expertise or tools instead of developing the skills or tools in house:

1. Your organization does not possess the skills that it needs to accomplish its goals.
2. Acquiring outside expertise or tools is more cost-effective than establishing them internally.

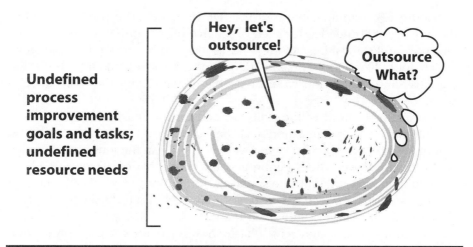

Figure 6.2   Scoping the Process Improvement Project before Outsourcing

However, before you can make the above decisions, you first have to know:

- The organization's business and process improvement goals
- What knowledge, skills, and tools your organization currently possesses

In other words, the organization needs to have some inkling of an idea of what constitutes the total project work before it can know what portion or aspects of that work to outsource. If the organization's leadership and process focus personnel can't see or scope process improvement even in terms of goals, as shown in Figure 6.2, then the scope of process improvement is an amorphous blob. In this situation, the organization doesn't want to simply ask a consultant to come in and "help us" or "do some stuff;" that is tantamount to handing them a blank check. However, the organization can hire outside expertise to provide some narrowly defined facilitation work to help the internal decision makers define the strategy and goals.

USING IDEAL FOR PROCESS ACQUISITION DECISIONS

SEI's IDEAL model[21] can give you a framework for determining the kinds of skills, experience, and knowledge your organization will need at different phases of CMMI-based process improvement.

Figure 6.3 is an adaptation of the IDEAL model. The picture identifies the primary skills needed for each of the five phases of IDEAL — Initiating, Diagnosing, Establishing, Acting, and Leveraging (or Learning). The icons which look like moon phases identify the relative amount of outside expertise an organization would reasonably acquire for the work in each phase.

For example, most of the skills needed for the Initiating Phase should be internal. The organization's vision, goals, and strategy should be determined by people who have a vested interest in the long-term viability of the organization. The thinking and the work that occurs in this phase is central to the organization's reason for existing and should not be determined by outsiders. A consultant may be hired to facilitate or coach these decisions, but not influence them.

In contrast, the Diagnosing Phase shows a relatively high proportion of the work being outsourced. Diagnosing usually involves some form of a method-driven characterization of the organization's current state of process implementation and discipline which frequently translates to performing an appraisal such as a CBA IPI or SCAMPI. The expert knowledge, skills, and experience required to perform process diagnostics are not typically indigenous to the organization and cost and schedule constraints typically don't allow for such skills to be internally grown at this stage of the journey.

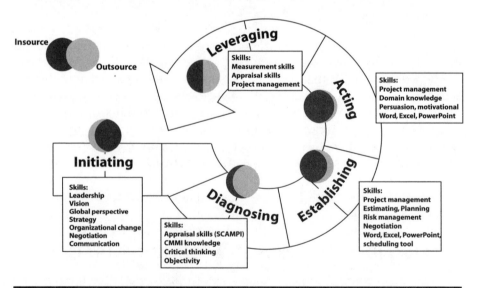

Figure 6.3   Using IDEAL as a Framework for Acquiring Process Expertise

In the Establishing Phase, the organization can use outside expertise to establish the process focus units or teams and to bring in process definition concepts that worked in other organizations. At the same time, the organization needs to start building its own internal process design and definition expertise during this phase.

The Acting Phase should belong almost entirely to the organization undergoing the change and there should be a limited need for external expertise. Most systems engineering organizations inherently possess the skills and ability required to pilot, train, implement, and manage new processes.

In Leveraging (learning), the organization can once again use some substantial CMMI expertise to coach and facilitate the process improvement lessons learned and to provide guidance for planning the next iteration of IDEAL in the organization.

This is just a framework — a guideline intended to provide you and others in your organization with a model for thinking about acquiring outside process expertise or tools. It is based on experience and observations and it is a model I've used successfully in planning my own consulting engagements. As always, you decide what is right for your organization and its process improvement work.

THE MATURING CLIENT–CONSULTANT RELATIONSHIP

Presumably, your organization's goals for CMMI-based process improvement include increasing its process capability in software or systems engineering and increasing the overall organizational maturity. These are certainly the implied goals if the organization is striving to achieve a CMMI maturity level. Extrapolating these goals out into the future, they would also logically include the organization increasing its process capability and maturity in continuous process improvement.

Translation: Over time the organization's dependence on the expertise of outside process or CMMI consultants should decrease as the organization's base of expertise in this area increases. And this is exactly the point at which your organization's goals and the goals of the consultant or vendor may not only differ, they may conflict. It is in the long-term best interest of your organization to gradually transfer outside process expertise and knowledge into the organization so that it gradually depends less on the outsiders for its process improvement success. However, it is usually in the best interest of the consultant or vendor to maintain a high level of dependency because it maintains their revenue. The consultancy can accomplish its revenue goal by ensuring its expertise and knowledge does not get transferred into your organization. See the problem?

What Usually Happens

The client and the consultant start out with competencies in different areas. The client has competency in their organization and business and the consultant does not. The consultant has competency in organizational change, CMM and CMMI, and process improvement which the client does not possess. As the relationship moves through time, each party should naturally transfer their competencies to the other. Thus, the client becomes more competent and capable of process improvement and the consultant becomes more effective in providing organization-specific solutions because of the increased organizational knowledge. However, as illustrated in Figure 6.4, this is not what often happens.

As shown in Figure 6.4, what often happens is that the consultancy is maintaining a high level of domain competency (CMMI) and effort. The client organization initially climbs a steep CMMI and process improvement learning and effort curve, but then their energy tapers off as process improvement gets out-prioritized by other deliverables and efforts and as they continue to rely on the consultant to do the work. The results of the relationship path shown in Figure 6.4 are great for the consultant, but not so good for the client.

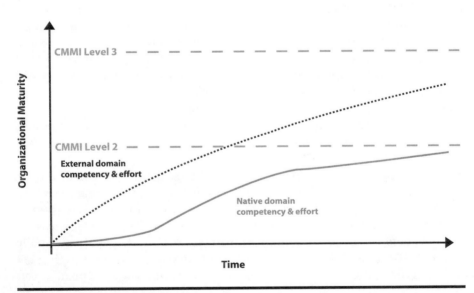

Figure 6.4 How the Client–Consultant Relationship Usually Matures

What Should Happen

Figure 6.5 shows how the client–consultant relationship should change over time to benefit the client.

In Figure 6.5, the consultant starts off with significantly greater domain knowledge (CMMI, process improvement) than the client organization which is, after all, presumably one of the reasons consultants are brought in to begin with. However, in order for the client organization to own the change and for continuous process improvements to become institutionalized, there needs to be gradual transfer of domain competency from the consultants to the client organization. As the client organization gains domain competency, the internal or native effort expended toward process improvement should also increase.

Of course, this presumes that there is a will and desire on the part of the consultant to transfer process improvement knowledge. It also assumes there is the will, ability, and availability of resources within the client organization to receive the transferring knowledge and skill.

The client–consultant relationship paths shown in Figure 6.5 are good for the client because the process improvement gradually becomes a core competency in the organization, which makes their success in this area less dependent on outside expertise. What many consultants don't realize is that this is also good for them and the consulting industry. Their revenue from a single contract may diminish over time, but their reputation and marketability for new clients or follow-on work or expanded work in a different domain with existing clients increases.

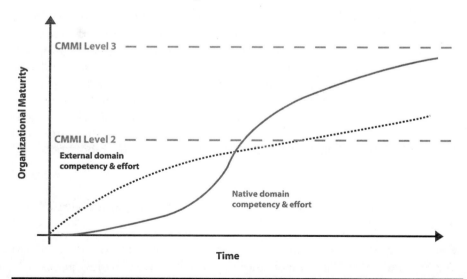

Figure 6.5   How the Client–Consultant Relationship Should Mature

DECISION CRITERIA FOR CONTRACTING CMMI OR PROCESS CONSULTING

Once your organization has made an informed decision to contract CMMI or process improvement expertise from a consultant, it's time to make an informed decision about which consultant to select and contract. After all, doesn't this procurement activity fall under the guidelines provided for SAM SP 1.2, Select suppliers? If not, why not?

Based on experiences and observations from people who have been on the client side and the vendor side of the consulting arrangement, here is a set of questions you can use to establish your organization's vendor selection criteria for hiring CMMI expertise.

What Is the Candidate Consultant Really Selling?

What is the candidate consultant really selling? Appraisals or appraisal planning and preparation? Maturity or capability levels? Training? Advice, based on what experience? Tools or process improvement automation? The bottom-line question the decision makers in your organization should ask and answer is, "Do the expertise, services, or products the consultant offers match your organization's needs?"

What Do the Candidate Consultants Say about ROI on Process Improvement?

What do the candidate consultants have to say about your organization's ROI or ROA on process improvement expenditure? Do they (without prompting) talk about measurable business benefits from process improvement? Do they ask you about the organization's business goals and talk about how process improvement can contribute to the achievement of those goals? How will they track, measure, and report improvement? Is your success and the consultant's success interdependent? The bottom-line question is, "Will your organization get the results it wants from the consulting arrangement?" (Corollary: Does your organization know the results it wants?)

Is Your Decision Based on the Consultant's Reputation or Verifiable Historical Performance?

Where have they worked and what have they done? Have they ever had "roll-the-sleeves-up" implementation experience or do they just know CMMI? Can they show historical performance measures such as actuals compared with plans? Do they give you referrals that check out? Are the people giving you the marketing pitch for the consultancy the same people who will be doing the work? If not, how can you be sure that the pitched capability will be the delivered capability? The bottom-line question is, "Is the vendor's reputation deserved?"

How Interested Is the Consultant in Learning about Your Organization's Current Process Capability or Organizational Maturity?

Do they propose a "solution" without first learning about your problems, goals, and priorities? Do they try to find out what you already have in place? Do they respect those things? Does the consultant's proposed "solution" fit your organization's philosophy, approach, and environment? How do you know? Is the consultancy marketing what amounts to a "slash-and-burn" approach to process improvement? Is that what you need? (Read Chapter 1 — News Flash! There Is a Level 1!)

DECISION CRITERIA FOR PURCHASING PROCESS TOOLS

Tools exist which can automate many software and system process tasks and activities:

- Requirements management and traceability
- Software and system size, cost, and schedule estimating
- Project scheduling
- Project tracking and reporting
- Risk analysis, planning, and managing
- Configuration and data management
- Measurement collection, analysis, and reporting
- Process asset definition and management

Tools can automate the work people do. Tools *cannot*:

- Improve leadership
- Keep anyone from making bad or irresponsible decisions
- Eliminate politics
- Fix a broken process (but they can make bad things happen faster)
- Motivate people
- Change the culture
- Relieve people of having to think

Often, the cost of the tool is just the tip of the iceberg of the total cost. Other costs include:

- Training and learning costs
- Integration with existing environment and platforms
- Initial loss of productivity during learning and transition
- Tailoring or customization
- Data migration from legacy systems
- Cost of running parallel systems or processes during transition

You can and should use the same vendor selection criteria you came up with for contracting a consultant as the criteria for selecting a process tool vendor. In addition to that criteria, the questions your organization should ask before buying a process or process improvement tool are:

- Does the tool do something you need or want? (Or is it just cool?)
- Would a lower cost, less elaborate solution satisfy your current and future requirements?
- How much are the unobvious labor costs: learning, tailoring and customizing, data migration, interfaces?
- How much expert labor could you buy for the cost of the tool?
- What will the tool do that can't be done just as easily and at a lower cost? At the current average CMMI consulting rates, you could not spend $200,000 on a tool and purchase about 1,300 hours of very good consulting. Which will benefit your organization more?

THE NUMBER ONE CONSULTANT OR VENDOR SELECTION CRITERIA: TRUST

When all is said and done, and your organization has established objective, fact-based criteria for selecting a CMMI or process improvement consultant or tool vendor, the ultimate selection criteria will come down to an answer to the question, "Who do you trust?"

If you're in the process business, you have undoubtedly heard that one of the biggest barriers to organizational change and CMM- or CMMI-based process improvement is that the stakeholders don't "walk the talk;" that is to say the people who have the most interest in the success of process improvement don't exhibit the behaviors or beliefs which they say are critical to success.

The same is often true of process consultants and vendors. It is not at all hard to find out if a potential consultant or vendor really understands CMMI and process discipline: if they understand it and believe in it, then you will be able to observe them "live and breathe" CMMI practices in their work. People who have really internalized CMMI incorporate its tenets into their daily lives almost at an unconscious (institutionalized) level.

So, when it comes down to your organization's leadership figuring out who they trust among CMMI consultants or tool vendors, they might consider figuring out who among the candidate consultants "walks the talk." Some questions you should try to answer are:

- Does the consultant gather requirements for work before proposing a solution? (REQM, RD)
- Does the consultant base estimates (e.g., proposals) on historical data using proven estimating methods? (PP, GP 2.2)
- Does the consultant develop realistic and manageable plans for the work they are going to perform? (PP, GP 2.2)
- Does the consultant monitor, track, and report progress and status against the approved plans? (PMC, GP 2.7, GP 2.8, GP 2.10)
- Do the consultants manage the versions or configurations of their own work product? (CM, GP 2.6)
- Does the consultant establish forms of objective reviews of their work against their own processes? (PMC, GP 2.9)
- Does the consultant manage its own risks? (PP, PMC, RSKM)
- Do the consultants include in their decisions and work the people and groups who will be affected? (IPM, IT, OEI, GP 2.7)
- Does the consultant collect and use measures of the work they perform to improve their service to your organization? (MA, GP 3.2)
- Do they ever ask you what they're doing well and what they need to improve? (MA, GP 2.7, GP 3.2)

DO'S AND DON'TS

The primary purpose of this chapter was to get you to view the acquisition of process consulting or tools through the perspective you presumably want everyone to have: a perspective based on informed decision making, managing by fact, and continuously improving process capability and maturity in *all* the organization's processes.

Here's the summary checklist of do's and don'ts to serve as a reminder of the major concepts presented in this chapter.

Do

- Recognize that acquiring process or CMMI expertise or tools is usually a very expensive proposition and that the decision makers in your organization should not take such decisions lightly.

- Decide to acquire process consulting for the right reasons such as cost, quality, or schedule.
- Select your process consultant or vendor based on objective selection criteria.

Don't

- Don't try to outsource your organization's process improvement; it cannot be done.
- Don't acquire process or CMMI expertise or tools just because it looks easy; make sure such decisions are based on your organization's requirements and plans for CMMI-based process improvement.
- Don't select consultants or vendors for the wrong reasons; only a miracle will prevent such a decision from becoming a very expensive and almost unrecoverable mistake.
- Don't think that just because consultants deliver "advice systems" or that process tool vendors deliver process tools that the principles and practices of CMMI don't apply to them. Hold all of your organization's vendors and consultants responsible for incorporating and upholding the concepts and ideals in which they ostensibly believe.

WHAT DID YOU LEARN? WHAT WILL YOU DO?

Now take the post-chapter quiz in Figure 6.6 and think about what you've learned and how some of your views toward CMMI-based process improvement have changed. Think about what you will do with the information you've learned (and how it makes you feel).

1. **Which of the following statements are true (may be more than one answer):**
    a. The practices found in the Supplier Agreement Management (SAM) Process Areas in the CMMI provide good guidelines for contracting a process consultant or tool.
    b. There really isn't any good way to select a CMMI consultant; you just have to roll the dice.
    c. The goals of process consultants are always the same as their client organization's goals.
    d. You should always try to hire consultants who guarantee that your organization will achieve a maturity level.

2. **True or False:** The type and amount of CMMI or process consulting your organization requires probably will vary based on the life-cycle phase of the process-improvement project.

3. **Process development, management, and implementation tools or systems can do which of the following for your organization?**
    a. Automate the management and traceability of system requirements.
    b. Help estimate project size, effort, and cost.
    c. Automate the storage, access, and configuration management of process documentation.
    d. Change the culture.
    e. Help with the collection, analysis, and reporting of measures.
    f. Ensure people make better decisions.

4. **The most important idea I learned from this chapter is:**

    _____

    _____

5. **I will apply this idea in my process improvement work by:**

    _____

    _____

Figure 6.6   Chapter 6: What Did You Learn? What Will You Do?

7

EFFECTIVE CHANGE LEADERSHIP FOR PROCESS IMPROVEMENT

Whenever someone comes to me for help, I listen very hard and ask myself, "What does this person really want — and what will they do to keep from getting it?"

— William Perry

ABOUT YOU, ABOUT THE QUOTE, AND ABOUT THE FUTURE

The ideas and messages in this chapter are an abstraction of an executive training course developed and delivered by Natural SPI called "Effective Change Leadership."[47] Perhaps you see yourself in the quote that starts this chapter and, if so, that's good because it means you're introspective and know that you can still learn something. If you don't relate to the quote at all, then you may want to keep reading about those "other" executives and senior managers, who are nothing at all like you.

When it comes to CMM or CMMI for process improvement, I know exactly what most of you are going to do and I know exactly what you're going to say. Most of you are managers, struggling with yourselves and struggling with the environment to become leaders. I know you. I've watched you, I've worked with you, I've tried to coach you, and I've tried to be your conscience and your guide. To get you to avoid the future which I see as fluid but which you have already accepted as fate, I've coaxed you, pleaded with you, bribed you, flattered you, groused at you, browbeat you, and I have given up on you, gotten exasperated, and walked away.

Whether you admit it or not, most of you want the same three things the rest of us want: (1) success (power and money), (2) respect (from yourself and others), and (3) both 1 and 2 to continue into your future. Regrettably, like the rest of us, you almost always sacrifice the third for the first two, never understanding that when you do so, you also lose the first two almost instantly. In your defense, you were probably so good in your area of expertise — software, engineering, sales, or accounting — that you were rewarded by being moved into management and leadership, an area in which you have little competence and for which you have been given no training, but for which you've been given tremendous responsibility. But one of your skills is winging it, so you cover the insecurity of not knowing what to do with hubris. In the short run, it works beautifully. I know, once and for a short time, it worked for me, too.

You are not exempt from the history or the statistically documented behavior patterns of your peer executives, you are not an exception, so here is what you're going to *say* to the organization when it comes to CMMI or process improvement:

- We must get to CMMI Level (pick your number, 1 to 5) to remain competitive.
- I want everyone to support the CMMI effort.
- Your job (or promotion or performance review or raise, you choose) depends on your contribution to CMMI (or getting the maturity level).
- The long-term viability of our enterprise relies on our process capability, not on our individual heroics.
- These are our processes and I expect all of you to follow them.
- CMMI is really important to all of us and our future.
- I expect all of you to embrace change and do things differently.

Those are the things you'll say. Here is what you're going to *do* or *not do*:

- When push comes to shove, when you are forced to choose between heroics and process to get the product out the door, you will choose heroics.
- You will reward the heroes, the people who work all night or all weekend to solve a customer problem. You will not reward the silent, humble engineer who puts quality first and prevents problems from ever getting to the customer.

- You will ask for status of the CMMI effort in terms of CMMI compliance. You will not ask for measures that indicate improvements in productivity, quality, cycle time, employee satisfaction, customer satisfaction, organizational learning, or other measures of operational excellence.
- You will not ask the process people to give you estimates for effort, cost, and schedule to achieve a CMMI maturity level. You will give them a target date that is tied to your bonus.
- You will tell the people who report to you that you want their support of the CMMI effort. You will not give them any incentive, positive or negative, for that support.
- You will not bother to learn the organizational processes yourself; they are for everyone else.
- You will not personally exhibit the change in behaviors you expect from others.

Alas, as so many who have gone before you, you will fail. Oh, you will get your maturity level/bonus/promotion/raise/praise/plaque on the wall, but make no mistake, you will have failed in leading your organization to change and grow. Of course, this isn't really you; it's those "other" leaders.

Here is the hope: You can follow the well-worn path just described or you can be the exception and break the cycle. The future, as it turns out, is really up to you.

OUR FUTURE IS YOURS ALONE[45]

Here's what the people who report to you want more than anything else: they want you to stop managing and start leading. They have been managed into oblivion, but they want to come back from the void. They want no more doublespeak from you; they will puke if one more time you excuse your decisions and actions by saying "it's political," and every time you say one thing but do another only further crushes their souls (and yours, too; only you may not think about it).

You want change in your organization? You want things to be different? You want a future that people buy into? You want success? Here's the best advice I can give you: Stop saying stuff and start doing things. Lose the adjectives and hyperbole from your vocabulary and start speaking and acting in imperative verbs. And there is no better place to start than with your organization's CMMI-based process improvement effort. The rest of this chapter tells you how.

THE DIFFERENCE BETWEEN MANAGING AND LEADING CHANGE

Again, I'm not the expert on leadership or organizational change. However, I have personally led organizational changes, and have been led through change by a rare few people whom I can justify calling "leaders." I can tell you this: despite what the literature might tell you, organizational change cannot be managed; it can only be led. When you think about it, the phrase "change management" is an oxymoron.

There is a difference between managing and leading and that difference is as pure and simple as the definitions for those two words. By definition, managing involves maintaining the status quo, not letting words, thoughts, or behaviors deviate too far from the acceptable "norm." In contrast, leading is willfully causing deviation from the status quo. Leading involves instigating the movement to get outside of the status quo; leading is the act of causing perturbation to a system that is static and in equilibrium so that it can morph into a better system. Management involves mitigating risk; leadership involves creating and taking risks. Management involves winning through complying with the rules; leadership often involves sometimes breaking the rules and frequently changing them.

In CMMI-based process improvement, your organization will be best served by a mix of leaders and managers. You will need managers who know how to use proven project and program management methods and tools to be in charge of the requirements, planning, design, training, and implementation of the processes and process improvement. You will also need leaders who inspire, encourage, and stand out in front of the organization as the first courageous examples of the new attitudes and behaviors.

The one thing we all need to come to grips with is the simple, observable fact that you can't find leaders by looking at the organizational chart. Positions, titles, and roles do not indicate leadership ability; sometimes just the opposite. Sometimes, the more important the title or the higher the position in the organizational chart, the more likely you'll find followers than leaders just because the rules of corporate America say that you move up by saying "yes" and by playing by the rules. Conversely, managers have people reporting to them, so the organization chart is a decent map for finding where managers live in the organization. Figure 7.1 illustrates the concept that leaders can be found in almost any strata, role, position, or title within the organization.

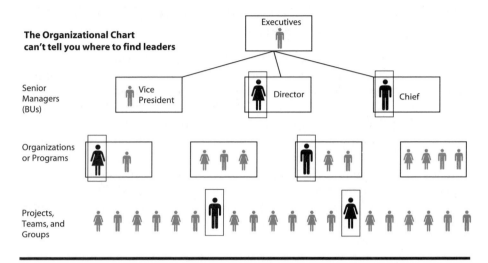

**The Organizational Chart
can't tell you where to find leaders**

Executives

Senior
Managers
(BUs)

Vice
President

Director

Chief

Organizations
or Programs

Projects,
Teams, and
Groups

Figure 7.1   Finding Where Leaders Live in the Organization

A VIEW OF CMMI FROM THE BOARD ROOM

Whether you actually are an executive level or senior level manager, you can easily assimilate what happens in the BOD (board of directors) meetings through corporate directives, policies, and memorandum and other artifacts, and acts which are issued forth from the BOD. You will also know directly what executives are thinking just by asking them.

At the executive or senior manager level, when it comes to things like CMM or CMMI process improvement, ISO, Six Sigma, etc., it's just one more thing to distract you from the latest Wall Street analyst's report on the company's stock performance or the waiting phone call from a customer CEO who wants personal attention on a problem with your product or service. Throughout the course of the day, the week, the month, and the year, the CMMI effort keeps getting moved to the bottom of your pile, often unconsciously. It's just not that loud or demanding at the moment; it can wait.

And wait it does until the day draws close when your organization wants to bid on a major contract, but the RFP (request for proposal) requires a CMMI process capability or organizational maturity level. Then and only then does it move up to the top of your pile, which is way too late.

DOING ALL THE RIGHT THINGS FOR ALL THE WRONG REASONS

Let's start by asking the one question so many people who report to you are afraid to ask: Why? Why is the organization "implementing CMMI?" Why is our organization trying to achieve a CMMI maturity level? It often costs midsize to large organizations hundreds of thousands or millions of dollars to implement CMMI-based process improvements or to achieve a maturity level; what will our organization get in return for this expenditure? Most people who report to you won't ask why because the obvious answer is that you told them to do it and that's good enough for them. However, you're at or near the top; you're the one making the decisions. Maybe you should think about why your organization is doing this CMMI stuff. Maybe some day, someone to whom you report will ask you "why?" You will probably want to have a good answer.

There are some very good reasons to use CMMI (and other models or bodies of knowledge) to improve the way people in your organization work. There are both quantitative reasons which are easily measurable and qualitative reasons which are not so easy to measure.

However, there are some really bad reasons to undertake CMMI-based process improvement. Process improvement, whether it's based on CMM, CMMI, ISO, Six Sigma, TQM, Critical Chain, or _____ usually represents a significant investment by the organization. Yet some of the common underlying reasons organizations often undertake these initiatives almost guarantee the organization will not be able to realize the return on its investment. Currently, many organizations start CMMI-based process improvement efforts for one or more of these three reasons:

1. The Holy Grail/Silver Bullet (HG/SB) syndrome
2. Corporate lemmingology
3. A golf course understanding of model-based process improvement

The Holy Grail/Silver Bullet Syndrome

The HG/SB syndrome is one of the common yet often unspoken reasons managers and executives initiate an undertaking as expensive as CMMI (see Figure 7.2). HG/SB is characterized by the following symptoms:

- A zealous belief that some magical combination of tools, processes, and models will solve all the organization's problems
- An obsession with throwing away resources on a crusade to find the HG/SB
- An immunity to most facts and data

Figure 7.2   The Holy Grail/Silver Bullet (HG/SB) Syndrome

There is no cure for the HG/SB syndrome except for executives to realize there is no HG/SB, never has there been one, never will there be one. Information and knowledge are the cure.

Corporate Lemmingology

Natural SPI (Figure 7.3) has coined this phrase to describe what many of us know as "keeping up with the neighbors." The symptoms of corporate lemmingology are:

- A pathological jealousy of other organizations (e.g., maturity level envy)
- An obsession with throwing away time and money to keep up with other organizations
- A culture of following (blindly), not leading
- An immunity to most facts and data

This is probably the most insidious of the bad reasons for taking on CMMI-based process improvement because its root cause is ego or fear: fear that your competitors will do something that helps their business while you get canned for not doing it too. The only cure for corporate lemmingology is for the leadership to stop following the pack, set their

Figure 7.3   Corporate Lemmingology

own course, and not worry when people say, "well, so-and-so is doing CMMI; why aren't you?" Courage is the cure.

A Golf-Course Mentality of CMMI-Based Process Improvement

This one is actually understandable (see Figure 7.4). No human being, not even Super-CEO, can possibly learn and know all the details of everything that is going on in his or her enterprise. This is one of the reasons organizations are hierarchical with multiple layers of management. At some volume of information flow, there is a critical mass of people needed by the organization to sort, sift, filter, and make sense of all that information. Regrettably and almost always inadvertently, the person at the top has to make some assumptions about something going on in the organization or simply has to accept and trust what she is being told by subordinates. This can lead to the allocation of valuable resources toward undefined or misunderstood goals and purposes. Such is often the case with CMM or CMMI. The executive only has the time to understand and believe that doing that "process improvement stuff" is good for the organization and let it go at that. What she doesn't realize is that she must get personally involved in leading that organizational change if it has any chance at all for success.

Figure 7.4   A Golf Course Understanding of Process Improvement

WHY PROCESS IMPROVEMENT IS SO DIFFICULT

All organizational change is hard and the change called CMMI-based process improvement is no exception. However, process improvement does have its own characteristics which make it a uniquely difficult organizational change, especially for leadership. Figure 7.5[46] illustrates one concept about the difficulty of success as it pertains to CMMI-based process improvement.

As Figure 7.5 illustrates, your chances as a leader to successfully implement change in your organization don't start off any better than one in four. Because success requires *both* appropriate decisions *and* effective implementation, the odds are against you from the outset. So, we'll get right to the point and give you examples of appropriate and inappropriate decisions and effective and ineffective implementation as they relate to CMMI-based process improvement.

Appropriate and Inappropriate Process Improvement Decisions

As an executive, there are really very few decisions you need to make regarding your organization's CMMI-based process improvement, but they are critical decisions that will determine the ultimate results and outcome. The critical decisions you need to make are:

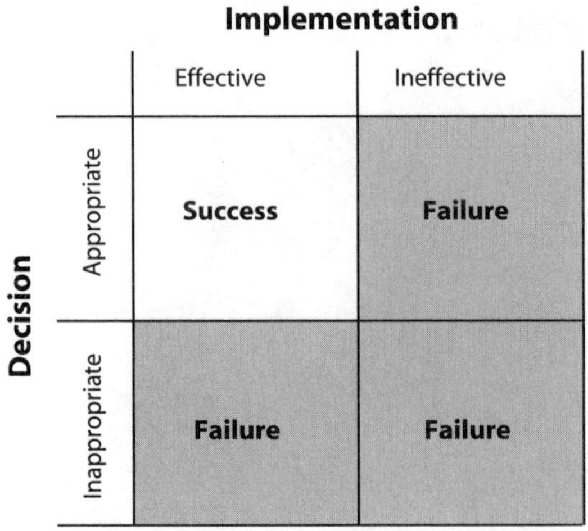

Implementation

Figure 7.5   Why Success Is So Difficult

Why?

Why is your organization engaging in model-based process improvement? What business goals do you expect to achieve or what problems do you expect to be resolved by use of CMMI? The section in this chapter, "Doing All the Right Things for All the Wrong Reasons" gives you some answers you'll want to stay away from. Chapter 4 — Process Improvement Strategies that Work tells you how to look at your organization from a systems-thinking perspective to understand that CMMI may not be the answer. Finally, Chapter 3 — Managing the Process Improvement Project — gives you techniques for determining how CMMI-based process improvement can address business goals and problems.

When?

This is actually the question you're probably going to try to answer first, especially if you've been given a directive from someone higher up the ladder than you. This is dangerous territory, so be very careful. This will be the first (and maybe last) chance you have to demonstrate your leadership in process improvement. Remember that a whole lot of process capability and organizational maturity is about fact-based estimating and planning. So, if you arbitrarily (i.e., without any factual basis) order your organization to achieve CMMI Level n by a certain date, you have already displaced trust with hypocrisy. In that one act, you will have clearly and overtly demonstrated that you have no intention of demonstrating the behaviors you expect to see in others, specifically making reasonable plans. The minute those words come out of your mouth, you have lost.

"Oh," you say, "I'm just giving them a stretch goal." Have you thought about what a "stretch goal" really is from someone else's perspective? How about this, a stretch goal is:

- Management's blatant lack of faith and trust in the ability of their people to accurately estimate and plan achievable work
- Management's unspoken goal and intention to not have to reward people for achieving goals given that "stretch goals," by definition, cannot be achieved

You've been taught to believe the myth that stretch goals are inspiring, that they get people to achieve more than they normally would. Nonsense. Stretch goals are demoralizing. Even the heroes will not try to meet the impossible stretch goals.

Who?

By "who?" we mean which parts of your organization really need to be involved in CMMI-based process improvement? What is the scope of the initiative? If you ask people on the help desk to implement all of CMMI, you may get a lot of people spending their time on HotJobs.com because only some small portion of CMMI can be applied to help-desk work. Before giving this answer, think about what CMMI really is — a model for operational excellence in system engineering and development — and then and only then decide which parts of your organization should undertake CMMI-based process improvement.

Which Questions Are Not Yours to Answer

One of the characteristics of a leader is knowing which questions and issues in which you should involve yourself and which business to stay out of. If you're an executive and you're getting involved in the detail "what" and "how" of process improvement, you haven't learned the role of an executive. You are paying very smart people a lot of money to figure out what to do and how to do it. They have read the books which you have not. They have taken the courses at SEI which you have not. They have attended the conferences which you have not. You have invested a lot of the organization's capital in building their knowledge and skill. If you now don't use that enhanced knowledge and skill, you have wasted the organization's money. Besides, you should have larger matters with which to concern yourself. Find your role and stick with it.

Effective and Ineffective Implementation

Let's get right to the point: If you're an executive, what business do you have in implementation? You are paid for your vision and ideas; you pay others to implement those visions and ideas. You may have come from the ranks of implementers, but now you are expected to lead, which doesn't usually involve sitting down at a keyboard and coding or writing a process description. This book is full of CMMI implementation concepts and techniques you can ask the people involved in your organization's process improvement effort to read.

There is one implementation concept not many people in your organization are likely to think about. Since it is a vision-level idea but affects the approach and implementation of CMMI, you're in the right role to address it. Figure 7.6 illustrates a proven approach regarding where and when to focus process improvement effort.

In Figure 7.6, the cross axes represent two relative scales. The X axis represents the relevance of process improvement work to addressing business goals or resolving business problems. The Y axis represents the relevance of process improvement work to satisfying CMMI process areas or practices. Many organizations start with the focus of their process improvement work high up on the Y axis but on the left end of the X axis. The work has high relevance to CMMI, but low relevance to their business objectives. The reason this happens is usually simply because

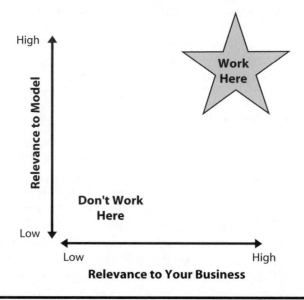

Figure 7.6 Where to Start CMMI-Based Process Improvement

the organization's leadership doesn't take the time to stop and figure out the relevance between CMMI-based process improvement and their business. Therein lies the executive's or senior manager's job. The most wasteful place to start with process improvement sometimes occurs in organizations in which there is very little comprehension of CMMI and a low understanding of the organization's business needs. In these cases, the organization will inadvertently start process improvement which has little or no relevance to either CMMI or their business. An example of this would be an IT help-desk organization writing procedures for the quality assurance of research papers — no perceptible model relevance, no perceptible business relevance, yet I've seen even worse.

THE LEADER'S ROLE IN CMMI PROCESS IMPROVEMENT

In CMMI-based process improvement, everyone involved falls into essentially one of three roles:

1. Sponsor (that would be you)
2. Change agent (sometimes you)
3. Target, which can be either a "victim" or participant (you, but only if you abdicate your role as a sponsor and leader)

Those of us who regularly attend process improvement conferences such as the SEPG conference have been subjected to an endless litany of "war stories" about executives and senior managers who don't really act as good sponsors of the organization's process improvement work. Fortunately for you, you're not at these conferences to hear it. The people who work for you go to these gatherings and they attribute your lack of sponsorship and process improvement leadership to lots of things, including:

■ You are malicious.
■ You are arrogant.
■ You are incompetent.

Based on the experience of my consulting firm working directly with executives and senior managers, they are wrong; you are none of those things. In fact, most of you actually want to be a good sponsor of change. You really want the CMMI effort to succeed and you want to have contributed to that success. The problem is that you really don't know what to do and your position as an executive or senior manager doesn't allow you to admit that you don't know what to do.

But before we get into specific techniques for leading process improvement in your organization, you should establish a fundamental understanding of your role in such an endeavor, the role of the sponsor.[46] In CMMI-based process improvement, you as the sponsor, do all of the following.

You own the change and communicate that ownership to everyone involved. The outward, public manifestation of this is standing in front of people and saying something like, "I own this change. You will get the credit for the success, but I will be responsible for failures because I am the leader. The buck stops here."

You select and empower qualified change agents to manage the change implementation. This means you guide the establishment of a process focus team within your organization and you allocate resources to that function. You should understand the skills and knowledge required for this work, find those people, recruit them, and give them incentive to be a part of the effort. If you need help understanding the skills and knowledge needed, read "Establishing the Process Improvement Project Team" in Chapter 3 — Managing the Process Improvement Project.

You establish an infrastructure to manage the change implementation. This means that you lead the establishment of new organizational structures, budgets, office space, and tools and equipment for the people involved in process improvement. You give them a home, not a tent.

You commit the resources that are required for success. Ah, here's the rub, money! You can't even begin to imagine how many of your peers in the executive and senior management ranks think that CMMI-based process improvement is free. They will actually ask people to "do that process stuff" on their own time and then they will be dismayed and confused when a year later no progress has been made toward the CMMI goal. As an engineer or lower level manager, you knew the axiom that the only things that are important in an organization are those things which get funded. Money talks and talk walks. It doesn't matter how many times you *say* that CMMI and process improvement are important, if it doesn't come with money, it won't come at all.

You keep the change in the consciousness of all subordinate reinforcing sponsors and targets. This means incorporating CMMI-based process work into your normal, everyday tasks. Don't treat it different from any other high priority, high visibility project or program that is underway. This effort is doing nothing less than delivering a process system that is going to ensure the long-term viability of the enterprise. Wouldn't you want a regular status on that project's progress and accomplishments?

You modify the reward system to encourage targets to change. It's this simple. Until process improvement shows up in the annual objectives, performance reviews, raises, promotions, and bonuses of people, or in public recognition, there will be no process improvement in your organization. Here's a precaution: there's a really ugly, cynical side to this story. If the only rewards, such as financial bonuses, go to you and other executive and senior managers for achieving CMMI goals, you will probably win in the short term. The organization will achieve its goals on the backs of people who go unrewarded and you will get your bonus. However, in the long term, you and the organization lose because eventually the talent you rely on for success gets wiser and looks for an exit.

You model any changes in behavior that you want others to adopt. This is the hardest role of all because it means finding the courage, intelligence, initiative, and integrity within yourself to change your own behaviors, long before you expect others to change. Regardless of what you've been trained to think, when the people who report to you are forced to choose between following your words and modeling your behavior, they will model your behavior every time. Why? Because it works for you and they want to be where you are. There are proven techniques for modifying your own behavior, but that is many other books and courses.

PROVEN TECHNIQUES FOR LEADING PROCESS IMPROVEMENT

As discussed earlier, on the whole, you and your executive peers are not malicious, arrogant, or incompetent, but you often don't have the information you need to successfully lead process improvement in your organization. And, given that you're the leader, you don't really want to admit to a lapse in your knowledge; you haven't arrived at your current position by focusing on your weaknesses, have you? So there, in the privacy of your office or your home, you can safely read this chapter and find out what you're supposed to do. You never have to admit that you learned it from another source.

Here are some proven techniques you can use to lead a successful CMMI-based process improvement program in your organization.

Things You Should Do

When communicating about the changes to come from process improvement, be brief. Describe where the organization is now, where it needs to go, and how it will get there. Identify who will implement the process improvements and who will be affected by these changes. Explain the criteria for success, the intended methods for verifying the success, and the related rewards. Identify key things that will *not* be changing. For example, let people know the processes, systems, rules, roles, etc., that won't be changing. This knowledge will provide the anchor they need to avoid the anxiety people suffer when "their whole world" is changing. Predict some of the negative aspects that the targets should anticipate.

Convey your commitment to the change. Tell people what changes they can expect to see in your behaviors, and tell them it's safe to let you know when they see a contradiction between your words and actions (and then actually make it safe).

When the people who report to you, especially program and project managers, give you status, don't accept phrases such as "we hope to ..." or "with luck, we'll ..." or "if everything goes right, then ..." These are the words used by people who don't manage their work and don't manage their risks using factual information. Gracefully and patiently teach them to come to you with factual plans and estimates that support their optimism. On the other hand, when things do fall apart, make it absolutely safe for people to give you bad news. Give them positive, public recognition for having the courage and integrity to deliver bad news that is based on reality and facts.

Don't accept status and progress that is not relative to a plan; it is meaningless except that it means a bunch of people are running around doing "stuff." Teach people to give you the status of their work against documented, approved plans and objectives because only then does status have any relevance.

Frequently ask people who report to you if they need any assistance, guidance, or help from you. They will get the clear message that you care about their work and its importance.

When people fail and then say to you, "but it will be different this time ..." ask them what fundamental shift has occurred in them, their approach, their work style, or the environment that would give you reason to believe that the next attempt at doing something the same way would yield different results.

Reward people who provide quantitative information that is verifiable. Discourage people whose vocabulary is replete with "sort of," "kind of," "maybe," and "almost."

Publicly reward and encourage people who find and remove system or product defects early in the life cycle.

When there is a crisis, turn first to the organization's processes for the solution.

Do your very best every single day to personally demonstrate and exhibit the behaviors you expect to see in those who follow you. We repeat this one for a reason.

Things You Should Not Do

Don't send conflicting messages. Don't allow your words to say one thing but your actions to speak something different. In the long run, the actions will win and the words will lose every time.

Stop rewarding the traditional "heroes," the people who work 24/7 to fix a problem or defect that has made it to the customer. Instead, gradually establish a new class of heroes, those who use standards and process discipline to prevent defects from going to the customer.

Don't punish people for taking calculated risks or for making mistakes. We don't learn nearly as much by doing things right as we learn from doing things wrong. People who have the courage to take risks and occasionally make a mistake are the people who will take the organization to new ideas, new products, new markets, and will save the organization from death by stagnation.

Don't punish people for saying the emperor has no clothes. If the organization has no tolerance for internal reflection and learning, then it has no chance of growing. If it's not growing, it is already dead; it's just too stupid to lie down.

8

PROCESS IMPROVEMENT MYTHS AND METHODOLOGIES

"But the Emperor has nothing on at all!" cried a little child.

— Hans Christian Andersen

ABOUT THIS CHAPTER

This chapter was actually the kernel of thought that started me down the path of writing this book. I was sitting around one day trying to make sense of all the things people kept telling me about CMM and process improvement. I guess I've always been somewhat of a skeptic, so I decided to go find out which of the hyperbole, myths, legends, and beliefs surrounding CMM and CMMI-based process improvement had any basis in fact or reality. As it turns out, many of the beliefs people in this industry were perpetuating were easily verified to be outright false or at least stretchy, elastic reality.

This chapter gives you a summary overview of CMMI and process improvement urban myths and what about them is mythical. Most of the details which debunk these myths can be found in other chapters in this book.

This chapter will probably affect you in one of three ways:

1. You will laugh along with me because you've heard the same things and knew them to be untrue and you're glad someone has finally put it in writing.

2. You will be shocked and appalled to find out that some of your most basic beliefs about CMM and CMMI-based process improvement are not factual.

3. You will be incensed, outraged, and angry at me for debunking myths that have served you so well over the years.

HOW MYTHS ARE BORN

One of the reasons CMM- and CMMI-based process improvement is so difficult is because of the urban myths which develop and then take on a life of their own. Often times, you find out that you'd have been better off if the engineering group had never even heard of either of these models.

How do myths originate? They originate the same way all other myths, rumors, and misdirected beliefs are born: they are born out of people having a little, but not quite enough, information, and filling in the gaps for themselves. Nature abhors a vacuum and people abhor lack of information and knowledge. When they don't have the means or the energy to fill in the missing pieces, they'll just make the stuff up. After a few repetitions, the myths become the new facts, the new reality.

Once the myth becomes institutionalized in the culture, it becomes very difficult for process focus people such as SEPG personnel or process managers to debunk them. I experienced a very interesting cultural phenomenon, in the form of language, when I changed jobs from Xerox to go work in the Application Services Division at CSC. Outside of CSC, the process improvement industry generally refers to maturity levels as "CMM levels," and software process improvement was called just that or the acronym "SPI." When I got to CSC, everything having anything to do with software process improvement or CMM was "SEI." In CSC, when organizations reached CMM Maturity Level 2, it was "SEI Level 2," the process improvement initiatives were "SEI programs," the CMM key practices were "SEI requirements," and people presumably knowledgeable about process improvement were "SEI experts." When I first started mentioning CMM, I got puzzled looks; people had no idea what I was talking about. It took me several months before I was able to get even a few key people in my organization to understand that SEI is a building and an organization at Carnegie Mellon® University and that it is CMM or CMMI which define maturity levels. Many people in CSC still won't understand the distinction until they read this book. Some who read it still won't believe it; such is the nature of people.

I learned an important ancillary lesson in this experience, one that's been known to sociologists for many years: language often defines peoples' beliefs, which in turn, defines their behaviors. Change the language and you'll change everything.

Sometimes, in spite of your efforts to bring facts into the daylight of the organization's consciousness, individuals will perpetuate myths for their benefit. For example, one of the popular myths I encountered at CSC was that there "were only 30 lead assessors in the world." Such myths are often perpetuated by people who perceive that the myth gives them more power or status.

CMMI AND PROCESS IMPLEMENTATION MYTHS

Over the years, I've collected some examples of ideas people have about CMM, CMMI, or process improvement. Take a few minutes to read through them. Think about each of the statements and how much truth or how much falsehood each represents. Then read the following sections to see if you, too, have been a victim of the process improvement myths:

- CMM or CMMI gives organizations requirements for developing successful processes.
- Having higher maturity levels ensures a software or systems organization will be successful.
- Before implementing CMM or CMMI, organizations are usually in total chaos.
- The primary and best reason for process improvement is to achieve maturity levels.
- CMM or CMMI will fix all your software development problems.
- Model-based process improvement doesn't affect what I do.
- Implementing process improvement based on CMMI is rocket science and only a few geniuses understand it.

So, before you can start your process improvement efforts using facts or natural, common sense approaches, let's first address some of the beliefs that could have you start off down the wrong track and waste a lot of your time.

Myth or Methodology: CMM or CMMI Gives Organizations Requirements for Developing Successful Processes

Pure fiction. SW-CMM and CMMI are models which provide organizations with guidelines for improving the software and systems engineering process. Neither SEI nor either of the models requires anyone to do anything (unless your organization happens to be a vendor, supplier, or subcontractor to SEI).

Furthermore, the practices defined in CMMI are not prescriptive; they help an organization determine "what" it takes to increase their process capability and organizational maturity, but by no means prescribe "how" to go about doing so.

This myth, if left to fester in an organization such that it becomes pervasive (institutionalized), can drive counterproductive behaviors. This misconception is what often leads managers to want to "implement CMM or CMMI" as opposed to improve their software or system delivery processes. This belief is one of the bases for people trying to achieve maturity levels instead of solving their business problems or achieving business goals.

Myth or Methodology: Having Higher Maturity Levels Ensures a Software Development Organization Will Be Successful

No, it does not. There is empirical data[42] which suggests that organizations at higher capability maturity levels have realized business benefits from their process improvement journey, but it's pretty hard to prove in organizations in which the only measurement collected is progress against achieving maturity levels.

There is also empirical data that suggests organizations can exhibit low systems delivery capability (e.g., cost and schedule overruns, poor product quality) even when they have been appraised at higher maturity levels.[43]

There are two reasons this belief is wrong. First, if an organization's only goal is maturity levels, that is exactly and only what it will get. (Actually, by luck, it may get other benefits, but it won't know what it doesn't measure.) If the organization's goal is to implement CMMI and get a maturity level, it is not very likely that the organization will bother to collect any measurements which would suggest the realization of business benefits such as performance improvement. If, on the other hand, the organization's goal is to use CMMI as a tool to fix problems, improve processes, and achieve business results, it is likely that it has planned its process improvement work to achieve those results, and will collect measurements to validate the investment in process improvement.

Myth or Methodology: Before Implementing CMM or CMMI, Organizations Are Usually in Total Chaos

If you haven't read Chapter 1 — News Flash! There Is a Level One!, then consider doing so. This belief is sometimes fact, but usually fiction. Have you ever heard someone in your organization use the phrase, "mud-sucking Level 1?" This is not only unnecessarily demeaning, its implications are usually not true.

To believe in this fiction is to not understand how CMM or CMMI evolved and to believe that there were no successful software or systems development organizations in the world before these models were published. Remember, CMM and CMMI are models that are abstractions of the best practices which evolved in software or systems organizations. The best practices and the successes preceded the model, not the other way around. In the postmodern business world, it would be unrealistic to think that any software or systems delivery organization would survive even one fiscal quarter without at least some structure, norms, and successful engineering and management practices in place. You can easily spot the organizations that are in total chaos; they don't live very long.

Myth or Methodology: The Primary and Best Reason for Process Improvement Is to Achieve Maturity Levels

This one should be obviously untrue, yet I'm sure you've seen behaviors that would suggest that it is fact. What is more likely to be true, based on my experiences, is that senior managers, higher up than you in the organizational chart, will personally benefit in the way of bonuses by making you and the organization achieve maturity levels. As a result of those personal motivations, achieving maturity levels becomes high priority for success.

As with most sustainable business ventures, the best reasons for an organization to throw hundreds of thousands or millions of dollars at process improvement is to eventually realize business benefits such as measurable defect prevention, schedule predictability, cost reduction, productivity improvement, employee satisfaction improvement, or competitive advantage. The eventual business benefits should exceed the investment.

If your organization is in the outsourcing business (e.g., SAIC, CSC, Keane), you'll often be told that in order for the company to continue winning contracts, it needs to achieve certain CMM or CMMI maturity levels. You'll hear that maturity levels give the company a distinct competitive advantage over its competition. This is true, but only in the short term; it is not a long-term sustainable reason for process improvement.

Here's why. Let's say you're told that the competing outsourcing company is promising its prospective customers that its organizations can achieve CMMI Level 3 in 24 months. So now your company's marketing people, who probably don't know what the acronym CMMI stands for, feel driven to make a better promise, say Level 3 in 18 months. Suddenly, you're hearing all around you the battle cry, "3 in 18." Once the contract is signed, someone will find a way to keep the promise, no matter how unrealistic it may seem. When fulfilling the contract (and thus big money) is at stake, people will do anything, including — yes — cheating, to get to those maturity levels.

Now, let's play this out a few years. Your company and the competition keep trying to trump each other in the CMMI maturity levels game. They promise "3 in 12," so you respond with "4 in 18." They promise "4 in 12," so now you're compelled to reach Maturity Level 5 in one year, and so on. At some point in the not too distant future, your company and its competitors are all able to promise the highest CMMI maturity level — 5 — in absurdly short periods of time. At that point in time, the maturity levels are no longer a competitive advantage, and they no longer contain the singular, albeit artificial, value they were given. The sales people will stop selling CMMI levels and move onto something else. The whole industry built up around model-based organizational maturity becomes irrelevant. Then, if you're in the process improvement business, you're in the really uncomfortable position of having invested a whole bunch of money into something which cannot show that it paid its own way, because all you were after and all you achieved were maturity levels. At that point in time, if you're part of your organization's process improvement initiative, you might consider updating your resume.

Myth or Methodology: CMM or CMMI Will Fix All Your Software and Systems Engineering Problems

If you think about it long enough (like about one minute), there's really no such thing as a "process problem" or even a "technical problem." Every problem we encounter in software or system development can be traced back to something someone said or didn't say or something someone did or didn't do. In other words, every problem, at its most root cause, is a people problem.

Even the very best procedures will not fix people problems. SW-CMM and CMMI don't even pretend to address people issues such as morale, learning curve, lack of flexibility, fear of change, incompetence, lack of skill, cultural norms, cluelessness, and on and on.

What's more, good procedures are a poor substitute for good leadership. It's really a cop-out to blame "the process," because that's much easier to do than figuring out a way to solve people problems.

What CMM and CMMI will do for you is help you understand a way to structure your software and systems engineering procedures so that they can be consistently performed by reasonably competent and motivated people. That is all. If your organization is hurting in ways which are not really related to process, see a doctor, see a counselor, hire a motivational speaker, but don't turn to one of these models for the answer.

Myth or Methodology: Model-Based Process Improvement Doesn't Affect What I Do

If you're a bug exterminator, this statement is probably true, and you're probably not reading this book. If you are involved in anyway in software or systems engineering, systems acquisition, or program and project management, model-based process improvement can have a dramatic impact on you and your work.

Myth or Methodology: CMM and CMMI Is Rocket Science and Only a Few Geniuses Understand It

There are a lot of people who would very much like for you to continue believing this, especially the process consultants charging $2500 dollars a day. Alas, it isn't true. Almost anyone with a high school reading ability and some experience in software or systems development can read and comprehend CMM or CMMI. Most major universities are now teaching these models as core curriculum in computer science tracks, so thousands of college graduates over the last five years have been coming out of school versed in at least the concepts of the model if not its practical implementation.

Applying the model in a sane way in your organization may take a little more comprehension. To apply the model in organizations, it will help to have a varied background in organizational change, sociology, communication, management, and systems engineering.

Here's one of the primary indicators of an organization that has — for whatever reasons — artificially elevated the difficulty of work in model-based process improvement: You'll walk into the organization and find out that everyone is unquestioningly doing what the "CMMI expert" or "SEI expert" tells them to do. When asked about CMMI, the "expert" will probably tell you to "not worry about CMMI, just do what I say." My advice to you: Run away.

APPRAISAL MYTHS

In the first part of this chapter, I've attempted to debunk some myths surrounding CMMI implementation in organizations, but wait, there's more! Since first becoming a SEI-authorized lead assessor in 1998 and then later an authorized SCAMPI lead, I think I've heard as many myths regarding maturity appraisals as I have about CMM or CMMI implementation. Some of them are downright bizarre.

Remember, as a change agent in your organization, you need to first understand why the CMM and the appraisal myths are being perpetuated.

The myths may be serving a vital purpose to someone in power in the organization. There are often better (read: safer) ways to get the emperor to put some clothes on than publicly stating that he's naked.

Take a minute to read through the following list of appraisal-related myths. Decide which ones you think are myths and which are fact-based beliefs. Then read the following sections to find out if what you believed about maturity appraisals is true.

Note: In some organizations, the word "audit" is used to refer to maturity appraisals or evaluations. Not only is audit technically inaccurate, it is also a negative, fear-inspiring word. If this is the word used in your organization, find out how that came to be, and try to get people to change the language. It's not beneficial to your process improvement efforts to have people trembling in fear of appraisals.

- When an organization passes an appraisal, it gets certified by SEI at a CMM maturity level.
- For an appraisal to be official, it needs to be led by someone from outside the company.
- You're not allowed to have people from your own organization on the appraisal team.
- If we can get an easy lead appraiser, we'll pass the assessment.
- We can pass the appraisal if we just get our people to give the right answers in the interviews.

Myth or Methodology: When an Organization Passes an Appraisal, It Gets Certified by SEI at a CMM Maturity Level

This one always amuses me. There have been individuals who I've told (and shown in print) at least three times that this is pure myth and yet, somehow, the next time those people mentioned appraisals, they talked about being "certified." They just simply want to believe.

But before we begin debunking the getting "certified" aspect of a maturity appraisal, let's first get rid of the pass or fail concept of assessments. Yes, ultimately, you may want an appraisal team to determine whether or not your organization has reached a certain maturity level. But if that's all the organization is after, then it's not mature, irrespective of what level rating comes out of the appraisal. The primary goal and the reason why an organization spends so much effort and money on an appraisal should be to determine its process strengths and weaknesses, which hopefully turn into action plans for continued improvement.

Now, back to the myth. So here it is again, try to get it this time — nobody certifies anybody at maturity levels. Not only does SEI not certify organizations at maturity levels, SEI emphatically and publicly states that it doesn't even validate assessment results that are reported to it. Here's SEI's statements that you'll find on its Web site with regard to assessment results that get reported to it:[49]

The terms SEI certified and CMM® certification are simply incorrect since there is no such thing.

The SEI does not certify organizations.

The SEI only licenses and authorizes lead appraisers to conduct appraisals.

The SEI or any other organization is a "certifying authority" of the results from an appraisal.

The SEI did not confirm the accuracy of the maturity levels reported in this list and has no intention of doing so.

Here's what actually happens: An appraisal team spends a week or more in your organization evaluating your software or system engineering practices against the practices and goals in CMMI. At the end of the appraisal, the appraisal team delivers a findings presentation. If it was documented in the approved appraisal plan that the appraisal would yield a maturity level rating, then the final findings presentation will include a statement indicating whether or not your organization has satisfied the goals of a particular maturity level. That's it. There's no plaque, no certificate from SEI, nothing of the sort. If you really want a certificate, drop me an e-mail and I'll print you one up for oh, let's say $50 dollars.

Myth or Methodology: For an Assessment to Be Official, It Needs to Be Led by Someone from outside the Company

This is myth. With regard to appraisals, you should probably eliminate the word "official" from your vocabulary, since you're not likely to find a consensus definition or meaning for it. At any rate, if you or your management really need to be able to call your appraisal "official," then just ensure that you plan and conduct an appraisal that is compliant with SEI's Appraisal Requirements for CMMI (ARC). In a SCAMPI Class A — which is the only class of CMMI appraisal that can yield a maturity level rating — the lead appraiser should not be from the "organizational unit" being appraised because this would represent a risk to objectivity. So, it really depends on how your organization has defined "organization" or "organizational unit" for the purposes of SCAMPI. If you're working in a large company, you may have process improvement or CMMI efforts going on in many subunits or organizations within the company. However, it is not likely that those separate efforts are coordinated at the corporate level to the extent you could perform a SCAMPI for the entire company in a single appraisal. Besides, the logistics of doing so would be a nightmare.

The reason many companies invest in sending people to SEI's Lead Appraiser training is so that they have internal resources to use to lead appraisals. Having internal people lead appraisals can save the organization anywhere from $20,000 to $40,000 per SCAMPI Class A appraisal. Sometimes, even though an organization may have access to internal lead appraisers, it may still hire someone from the outside because the management perceives a need for a higher level of experience and expertise than that possessed by its internal people.

Myth or Methodology: You're Not Allowed to Have People from Your Own Organization on the Assessment Team

Not true. In fact, the SCAMPI methodology requires that some portion of your appraisal team membership be people from within the organization being assessed. These internal team members serve a vital purpose in the appraisal. They serve as guides by helping the entire team understand how the organization's work maps to CMMI. Without internal people on the team, appraisals would take much more time and cost more money on the average than they do.

Appendix A

REFERENCES

1. Paul, Mark C. et al., *The Capability Maturity Model, Guidelines for Improving the Software Process,* Addison-Wesley, 1995.
2. The Capability Maturity Model Integration^{SM}— SE/SW/IPPD/SS, V1.1, CMU/SEI-2000-TR-030, Carnegie Mellon University, March 2002.
3. Standard CMMI^{SM} Appraisal Method for Process Improvement (SCAMPI^{SM}), Version 1.1: Method Definition Document (MDD), CMU/SEI-2001-HB-001, Carnegie Mellon University, December 2001.
4. Reiner, Rob, *This Is Spinal Tap,* Metro-Goldwyn-Mayer. 1984. Script downloaded from http://www.krug.org/scripts/tap.html.
5. Dunaway, D. and Masters, S., CMM-Based Appraisal for Internal Process Improvement (CBA IPI): Method Description (CMU/SEI-96-TR-007), Carnegie Mellon University, April 1996.
6. Tegmark, Max, Parallel Universes, *Scientific American,* May 2003.
7. Goldratt, Eliyahu, *Critical Chain,* North River Press Publishing Corporation, April 1997.
8. Johnson, S. and Blanchard, K., *Who Moved My Cheese?* Putnam Publishing Group, 1998.
9. International Standard 9001, Version 2000-12-15 (ISO 9001:2000E), ISO 2000.
10. Baldrige National Quality Program: go to http://www.quality.nist.gov/.
11. Pande, P. et al., *The Six Sigma Way: How GE, Motorola, and Other Top Companies are Honing Their Performance,* McGraw-Hill Trade, April 2000.
12. Tingey, Michael O., *Comparing ISO 9000, Malcolm Baldrige and the SEI CMM for Software,* Prentice Hall, 1997.
13. ISO to CMMI-SE/SW/IPPD, V1.1, SEI, Carnegie Mellon University, 2001.
14. Rico, David F., Clause-By-Clause Comparison of ISO 9001:2000 and CMMI, 2002. Go to www.davidfrico.com for more information.
15. Mutafelija, Boris, *ISO 9001:2000 to CMMI Mapping,* Hughes Network Systems, 2002.
16. The Project Management Body of Knowledge (PMBOK) is an online repository of project management information. This repository is maintained by the Project Management Institute (PMI). For more information, go to www.pmbok.com.

17. West, M. and Byrnes, S., CMMI and PMBOK Synergy: Leveraging Process Platforms for Improvement, presentation delivered at the 2003 National Defense Industrial Association (NDIA) CMMI Technology Conference and Users Group, Denver, CO, November 2003.

18. Practical Software and Systems Measurement is an organization devoted to developing, establishing, and communicating methods and tools for measuring software and systems work products and processes. For more information, go to www.psmsc.com.

19. Basili, Victor R., Software Modeling and Measurement: The Goal/Question/Metric Paradigm, Technical Report, CS-TR-2956, Department of Computer Science, University of Maryland, College Park, MD 20742, September 1992.

20. Kaplan, R. and Norton, D., *Translating Strategy into Action: The Balanced Scorecard,* Harvard Business School Press, 1996.

21. McFeeley, B., IDEAL^SM: A User's Guide for Software Process Improvement, Technical Report CMU/SEI-96-HB-001, February 1996, Carnegie Mellon University.

22. Adapted from Senge et al., *The Fifth Discipline Fieldbook: Strategies and Tools for Building a Learning Organization,* Currency-Doubleday, 1994.

23. Amputating Assets: Companies That Slash Jobs Often End Up with More Problems Than Profits, *US News and World Report,* May 4, 1992.

24. Adapted from Bergquist, William, *The Post Modern Organization: Mastering the Art of Irreversible Change,* Jossey-Bass, Inc., 1993.

25. An unsolicited advertisement from Castilon Consulting LLC dated July 15, 2003 sells, "Level 0 to 3 in less than 6 months." The brochure cites a case study in which Castilon Consulting helped an IT organization rapidly achieve CMM Level 3. In the "Benefits" section of the advertisement, Castilon identifies 13 "benefits" that can be realized from their approach. However, nowhere in the entire brochure do they mention anything about their clients achieving even a single business improvement (e.g., ROI, ROA, productivity, product or service quality, customer satisfaction, employee satisfaction) from the Castilon approach. Apparently, the rapid achievement of a CMM maturity level is the benefit. They're probably doing okay since there seems to be quite the market for maturity levels.

26. The Maturity Profile is a report periodically published by the Software Engineering Institute (SEI). This document provides information and statistics from the results of CMM and CMMI appraisals reported to SEI. For more information, go to www.sei.cmu.edu.

27. In 2002 and 2003, IBM ran a series of television advertisements which poked fun at the belief there could somehow exist a single, magical, universal integrated solution for any and all business problems and goals. The ads were powerful in their ironic comedy.

28. Bogan, Christopher E. and English, Michael J., *Benchmarking For Best Practices: Winning Through Innovative Adaptation,* R.R. Donnelley & Sons, 1994.

29. As of August 2003, a search on Amazon.com yielded five titles related to CMMI and its implementation. Reference www.amazon.com.

30. For more information on standards and work products available from IEEE, go to http://www.ieee.org/portal/index.jsp.

31. The quarterly Maturity Profile published by SEI on its Web site shows the number of organizations that have reported their CMM assessment results. The data is parsed in numerous ways including by industry sector (SIC), geographical area, and organization size.

32. Gartner Group definition for "best practices."

33. West, Michael, Project Management Best Practices, delivered to the California Management Accountants (CMA) in Solvang, CA on August 17, 2001.

34. WITNESS is a process simulation product of Lanner Corporation. For more information, go to http://www.lanner.com/corporate/.

35. SIMUL8 is a process simulation product of SIMUL8 Corporation. For more information, go to http://www.simul8.com/.

36. Nohria, N., Joyce, W., and Roberson, B., What Really Works, *Harvard Business Review*, July 2003.

37. Adapted from Tushman & Anderson, Technological Discontinuities and Organizational Environments, *Administrative Science Quarterly*, 14, 311-347, 1986.

38. Wiegers, K., Software Process Improvement: Ten Traps to Avoid. Go to www.processimpact.com.

39. Elsbach, K., How to Pitch a Brilliant Idea, *Harvard Business Review*, September 2003. The quote is a paraphrase attributed to Robert Goizueta, Coca-Cola's CEO in 1985, the period of the disastrous introduction of a new formula for Coke.

40. West, M., Acquiring Process Expertise and Tools: A Fact-based Methodology, presentation first delivered at NDIA CMMI Technology Conference and Users Group, Denver, CO, November 2003.

41. Based on Natural SPI proprietary data.

42. Some of the best data collected to date on business benefits resulting from software process improvement were documented in a case study performed by The Boeing Company on one of its subsidiaries, BCS Richland. This information was presented by Boeing's John Vu at the 19th Annual International Software Process Engineering Conference in Seattle, WA.

43. At the 2002 SEPG Conference in Phoenix, AZ, Tom Laux, then Deputy Secretary for Acquisiton for the Navy, gave a keynote address. The crux of the presentation was that Naval Aviation (NAVAIR) had learned a hard lesson. In the late 1990s, NAVAIR required all contractors to be CMM Level 3. Well, NAVAIR got contractors which had been appraised at CMM Level 3, but many of them were still demonstrating poor systems engineering and delivery performance, they were still overrunning cost and schedule, and delivering defective systems. As a result, NAVAIR changed its policy and began requiring prospective contractors to show or demonstrate systems delivery performance.

44. This information and subsequent information in this section is informed by a presentation developed and delivered by Dr. Rick Hefner and Leitha Purcell, both of Northrop Grumman Mission Systems. The presentation, Software Applications of Six Sigma, was delivered to the Los Angeles Software Process Improvement Network (SPIN) on January 29, 2003.

45. Taylor, Jim and Wacker, Watts with Means, Howard, *The Visionary's Handbook: Nine Paradoxes That Will Shape the Future of Your Business,* HarperBusiness, 2001.

46. This reference and subsequent references are adapted from SEI's course, Managing Technological Change, which adapted materials developed by Implementation Management Associates, Golden, CO.

47. Voight, D. and West, M., *Effective Change Leadership (ECL)*, Natural Systems Process Improvement, Inc., 2003.

48. Byrnes, P. and Phillips, M., *Software Capability Evaluation (SCE)*, Version 3.0, CMU/SEI-96-TR-002, Software Engineering Institute, Carnegie Mellon University, April 1996.

49. From SEI's Web site. Go to http://seir.sei.cmu.edu/pml/for more information.

50. Segal, Peter, *Anger Management*, Sony Pictures, 2003.

INDEX

Note: Italicized page numbers refer to illustrations, notes, and tables

A

B

C

Q